Titanium-Based Alloys for Biomedical Applications

Petrică VIZUREANU and Mădălina Simona BĂLȚATU

Published by **Materials Research Forum LLC**
Millersville, PA 17551, USA

Published as part of the book series
Materials Research Foundations
Volume 74 (2020)
ISSN 2471-8890 (Print)
ISSN 2471-8904 (Online)

Print ISBN 978-1-64490-078-9
ePDF ISBN 978-1-64490-079-6

Distributed worldwide by

Materials Research Forum LLC
105 Springdale Lane
Millersville, PA 17551
USA
http://www.mrforum.com

Printed in the United States of America
10 9 8 7 6 5 4 3 2 1

Table of Contents

CHAPTER I: Introduction and Opportunities

At the global level, there is a continuous concern for the research and for the development of alloys for medical and biomedical applications. Thus, it is desired to improve both the classical implant manufacturing technologies and the biomaterial synthesis technologies from which they are being executed, with the ultimate goal of promoting a new generation of multifunctional implants with long-term performance.

Metallic biomaterials are used in various applications: orthopaedic, dental and cardiovascular. As biocompatible metallic biomaterials, various materials are used, such as stainless steels, Co-based alloys and Ti-based alloys. Each class of metallic biomaterials has advantages and disadvantages and the use is made according to the properties imposed by the functional requirements of an implant.

Titanium and its alloys are of the greatest interest in medical applications because they have important characteristics required for implant materials, namely, good mechanical properties (lower elasticity modulus than stainless steel or CoCr alloys, fatigue strength, corrosion resistance) and high biocompatibility. These materials have multiple uses in orthopedics, dentistry, maxillo-facial surgery and cardiovascular surgery.

Contributions for research and practice are presented taking into considerations numerous pieces of information on metallic biomaterials, especially titanium alloys: nevertheless, these bits of data are focused on the recent developments and applications of biomaterials. The novelty of the book is to promote new alloys and the characterization of them through laboratory tests in order to identify all their properties. Alloys obtained and characterized in the book are Ti-based with different concentrations of non-toxic chemical elements (Mo, Zr, Ta, Si, Nb, etc.).

The impact of the book is important because it highlights health/medical applications which can be improved with titanium biomaterials with non allergenic elements to meet functional requirements and to remove the disadvantages of classical alloys used as biomaterials in human tissue.

This book is unique because it contains new recipes of titanium alloys with excellent properties: low density, good corrosion resistance, low elasticity and high biocompatibility and its very good tolerance in the human body, as compared to that of stainless steel or CoCr alloys.

Having taken into consideration from the literature on metal alloys, we have developed a new TiMoZrTa (TMZT) system, which removes the disadvantages of Co-Cr alloys and stainless steels. The new system version contains different alloying percentages of non-toxic elements (Mo, Zr, Ta) in order, to improve mechanical properties (mechanical strength, modulus of elasticity), corrosion resistance and biocompatibility.

Biomaterials have to meet certain requirements and also, they have to take into account some important characteristics. Properties of interest in the use of these alloys are: mechanical properties, corrosion resistance and biocompatibility.

In order to identify the biocompatible character of the developed alloys, it is necessary to perform specific laboratory tests for structural characterization, mechanical characterization, corrosion resistance assessment and cytotoxicity assessment. Carrying out these types of investigations will provide sufficient information to compare alloys obtained with other types of biomaterials to assign a suitable medical application.

The book contains eight chapters.

The first chapter entitled *"Introduction and opportunities"* presents introductory concepts, about Ti-based alloy as metallic biocompatible materials.

In the second chapter entitled *"The state-of-the-art on biomaterials used in the human body"* is presented an actual state-of-the-art, of titanium alloys with appropriate characteristics for use in the medical field. Diversification of these types of materials requires a surface functionality and their specialization depends on the following goal.

The third chapter entitled *"Objectives and methodology of experimental research"*, presents the methods of analysis that have been proposed, as well as the equipment used, which are in concordance with the technical standards and norms for metallic alloys with applications in medicine.

Chapter four entitled *"Obtaining of new Ti-based alloys for medical applications"*, presents the development of new original alloys, obtained on an arc vacuum furnace. Some advantages of using this equipment are: very high melting temperatures can be achieved, and alloys with uniform composition can be created, through repeated remelting.

The fifth chapter entitled *"Investigation of Ti-based regarding structural, mechanical, electrochemical characterization and biocompatibility assessment"*, presents a complete structural characterization of new alloys, physico-mechanical properties, electrochemical behaviour and biocompatibility aspects.

The sixth chapter entitled *"Optimization of future Ti-based alloys biomaterials"* presents an analyse about the content of the book and the variants that could be optimized in the future, considering the toxicity of some chemistry elements that has recently been demonstrated (Al, V, etc.).

The seventh chapter entitled *"Destination of Ti-based alloys used in human body"*, highlights the main applications of biomaterials used in the human body (medicine, cardiology and dentistry), like: cardiovascular implants, prosthesis joints, dental implants, spinal implants, orthopaedic plates and screws, etc.

The last chapter entitled *"Exploitation and trends"* contains future opportunities to develop Ti-based alloy in medical applications by adding other biocompatible chemical elements or functionalizing the surfaces of the alloys already obtained by coatings.

Materials Research Forum LLC
https://doi.org/10.21741/9781644900796

CHAPTER II: The State-of-the-Art on Biomaterials used in the Human Body

II.1. General Considerations on Metallic Biomaterials

By the term biomaterial is meant any entity of material that can be used in contact, endogenous or exogenous, with the tissue of a living animal for the purpose of its regeneration, replacement, therapeutic, diagnostic or monitoring [1].

Biomaterials play an important role in many aspects of the contemporary medical field through considerable progress over time. These are products of organic or inorganic nature that have found various uses: prostheses or implants in biological tissues in humans or animals. In order to obtain these materials used for biological purposes, metal, ceramic, organic polymers, etc. are used [2-7].

Applications of biomaterials in the field of medicine are primarily due to the requirements of medical practice, but also to the continuous evolution of the sciences. A permanent correlation of research in the fields of chemistry, biology, engineering, and medicine drives the science of biomaterials to obtain new materials that can solve the many existing medical problems under current requirements [7].

The use of materials for the reconstruction of tissues in the human body has its beginnings a few thousand years ago, but clinically important advances have been achieved in the last century. The first replacements of affected tissues date back more than 2000 years and are related to Aztec, Chinese and Roman civilization [8].

Despite the large number of metals and alloys produced in industry, only a few of them are biocompatible and successfully used for the long term in the human body. Ludwigson [9] differentiated the use of various materials in medical implants and dental applications in three chronological generations:

1.   The first generation covers the period 1860-1870, respectively until the aseptic surgical techniques were applied;

2.   The second generation is between 1870-1925 and corresponds to the period of great achievements and discoveries in the field of science and technology;

3.   The third generation, the most modern one, ranges from 1925 to the present and coincides with a new stage in the development of materials technology used as biomaterials [8, 10].

Table II.1. History of materials used for reconstruction of tissues [8, 10].

Time period	Material	References
Befor of 500 IC year	Gold, wood, ivory	Archaeological evidence from China, Egipt, Italia
1565	Gold	Pretorius
1666	Gold, ivory	Fabricius
1775	Brass	Pujal
1827	Platinum	Levert
1829	Silver	Rodgers
1860	Aseptic materials	Lister
1902	Tantalum	Lambotle
1912	Stainless steel	Sherman
1924	Stellit (Co / Cr Alloy)	Xierold
1936	Cobalt alloys	Venable
1940	Titanium	Leventhal
1957	Poly (methyl methacrylate) - PMMA, Polyethylene-PE, Polyamide-Nylon	------
1960	polyurethanes	-------
1962	Ceramic materials	Smith
1960	Cellulose, natural rubber, graphite	-------
1969	Politetrafluoroethylene (Teflon)	Gore
1970	Hydroxyapatite, tricalcium phosphate	-------
1970	Composite materials, hydrogels	-------
1999-2000	Nanomaterials	-------
2000-present	Biodegradable materials and multicomponent alloys (binary, ternary, quaternary)	-------

As highlighted in Table II.1, where a review of materials used in implantology is presented, most biomaterials have been developed, improved and used in the last century.

There are several criteria for classifying the materials from which the implants are made, but a frequent classification of biomaterials is made by their physico-chemical nature, as it follows: metallic, ceramic, polymeric and composite biomaterials. Table II.2 lists the main types of biomaterials, together with their advantages and disadvantages [10-13].

The use of biomaterials is based on properties, but also on the functional requirements imposed on implants.

The common element of all biomaterials is their biocompatibility with human tissue, so that the response to the interaction between the implant and human tissue is positive without causing adverse reactions [13, 14].

Table II.2. The main types of biomaterials [12].

CLASSIFICATION		ADVANTAGES	DISADVANTAGES
Metallic biomaterials			
Pure metals	Au, Ag, Pt, Ti, Ta	High mechanical resistance, machinability through deformation, toughness	Corrodes in the environment of the human body
Alloys	Stainless steels Ti-based (Ti-Al-V, Ti-Al-Fe,Ti-Al-Nb) Cobalt based (Co-Cr-Mo, Co-Cr) Dental amalgam with shape memory (Ni-Ti)		
Biodegradable alloys	Iron-based Magnesium based		
Ceramic biomaterials			
Ceramic	Based on oxides (Al2O3, ZrO2) Based on carbides and nitrides (TiC, TiN) Variants of carbonate (vitreous, pyrolytic)	Resistance to high compression, bioactivity, chemical inertia	Fragility, difficult production
Bioactive ceramics	Based on calcium phosphate (hydroxyapatite) Based on other calcium salts (carbonates, sulphates, aluminates)		
Biomaterials based on synthetic polymers			
Elastomers Plastic materials	Silicones, polyurethanes Thermosetting (epoxy resins, triazine) Thermoplastic (polyethylene)	Easy to fabricate, resilient, relatively low cost, adaptability of properties	It deforms over time, it can degrade in the biological environment, insufficient mechanical strength
Bioresorbable	Polyglycolic acid, polylactic acid		
Synthetic composite biomaterials			
Organo-organic type Mineral-mineral type Organo-mineral type		Adaptation of properties and dimensions by application	Difficult production

The first metal alloys used to make implants used in orthopedics were stainless steel alloys with vanadium and cobalt-chromium alloys. After 1940 titanium alloys (pure commercial titanium) were introduced. The first ones who obtained titanium dental implants were Linkow (1968), Branemark (1969) and Hofmann (1985), who, however, also investigated a titanium alloy: Ti6Al4V [15-19].

Nowadays, 316L stainless steel is used for medical implants due to its corrosion resistance, along with a wide range of other physical and mechanical properties. A disadvantage of austenitic stainless steels is their high sensitivity to corrosion and pitting [13, 19-22].

CoCr alloys have a wide spread in medical applications due to advantageous properties and relatively low prices. A disadvantage for Co-based alloys is the modulus of elasticity (220 ~ 234 GPa), much higher than that of stainless steels and titanium alloys [19, 22-24].

The use of titanium in implantology is due to its excellent properties: low density, good corrosion resistance, low modulus of elasticity and high biocompatibility. Frequent use in medical applications is due to its tolerance in the human body, compared to that of stainless steel or CoCr alloys [25].

Changing surfaces through chemical processes or introducing biomolecules to control the process of attaching living cells and protein assimilation is an important option for improving biocompatibility (inorganic biomaterials). Many of the metallic elements in the alloys used in medical applications are needed in the human body as micronutrients. Some of them help in the growth, development and physiology of the body, but are toxic in quantities greater than required (Fe, Cu, Mn, Zn, Se, Co, Mo, Cr) [16, 21].

The usual materials currently found in medical applications are classical metal alloys: titanium alloys (TiAlV), cobalt alloys (CoCrMo) and stainless steels. Research in the field of biomaterials over the last decade has helped to improve properties by introducing non-toxic elements, thereby improving surface, mechanical, corrosion resistance, biocompatibility, etc., with the aim of replacing existing ones [26].

II.2. Biomaterials Requirements in Medical Applications

Designing and selecting biomaterials consists in choosing the correct material compositional limits so that unwanted reactions do not occur through human tissue implant interaction [26].

The design of a material for use in medical applications takes into account biocompatibility, mechanical properties and corrosion resistance. In addition to the aforementioned aspects, the surgical technique applied to implantation and, last but not least, the patient's states of health are added to its functionality.

The biocompatibility of a material can be defined in the sense that it produces desired or tolerated reactions in a living organism. The metals, in contact with the biological body, give complex effects, producing a series of biological reactions depending on the concentration, the exposure time, etc. According to the biological interaction mode, the metals are divided into [5, 24-27]:

Metallic elements required in very small concentrations for the living organism called essential elements, including cobalt, manganese, zinc, magnesium, sodium, potassium, etc.; Elements that produce toxic effects for the body if present at higher concentrations, the cytotoxicity effect being demonstrated by the cell culture system, such as arsenic, cobalt, nickel. etc.;

Metals with allergic potential (nickel, cobalt and chrome) are considered to be highly allergic to the body.

The overall strategy for testing biomaterials involves assessing them in two ways [22]:

(i) "in vitro" evaluation performed on cell cultures or in the blood;

(ii) "in vivo" assessment on animals.

Depending on the material-body interactions and the possible reactions occurring, the biocompatibility of the material can be assessed. The sequence of tests applied in the evaluation of a biomaterial is presented in Table II.3 [5, 24-27].

Table II.3. Sequence of tests applied in the assessment of a new biomaterial [22].

"In vitro" tests	cytotoxicity - culture cells Hemolysis regarding the human or animal blood mutagenicity - human or other mammalian cells, bacteria
"In vivo" tests	acute systemic toxicity of the mouse sensitivity - guinea pig pyrogenicity - rabbit intracutaneous irritation - rabbit intramuscular implant - rabbit hemocompatibility - rat, dog, ... long-term implant - rat, pig, ...

The mechanical properties of biomaterials are dependent on the chemical composition, structure and function of the implant in the living organism, and have corresponding properties similar to those of the tissue they are replacing.

Mechanical properties to be considered in the evaluation of a biomaterial are: hardness, the longitudinal elastic modulus, the tensile strength and elongation. The material's response to repeated cyclical stresses is determined by its fatigue strength, one of the mechanical properties that provide long-term resistance to the implant. Alloys used for orthopedic implants must have a longitudinal elastic modulus similar or similar to bone that varies between 4-30 GPa, depending on the bone type and the direction of measurement [27-29].

Corrosion of implants results in the release of metal ions in the body, considered responsible for producing allergies and toxic reactions in the human body. For most alloys used in implanted medical applications, the rate of corrosion is mainly dependent on the properties and stability of the passive film formed at its surface. The quality of passive film protection is dependent on its ability to resist to chemical / electrochemical degradation caused by the human body's environment, as well as its ability to repassivate quickly enough to avoid initiating attack of the metal substrate [30].

The overall response of a living organism to an implant is governed by a number of factors that determine whether the implant is accepted or not and can fully perform its function.

To make medical applications from a selected material, several factors have to be considered. (Table II.4) [31].

Table II.4. Factors taken into account in the selection of biomaterials.

FACTORS	economic	raw material costs, processing costs
	mechanic	tensile strength, compressive strength, tear strength, shock resistance, fatigue strength
	electric	electrical resistance, electromechanical compatibility
	chemical	humidity, recycling capacity, pollution, corrosion resistance
	biological	toxic toxicity, toxic interaction, biocompatibility with human tissue
	thermal	thermal expansion, thermal stability
	surface	wear, rubbing, finishing
	aesthetic	economic aspect, color, ergonomics, visual clarity
	performance	achievement of the goal pursued, user tolerance
	research	future use, novelty

When selecting a particular implant material, also it must be taken of the health of the implantation area, both in an uninfected and healthy context, and in a morphologically relevant tissue quality of implantation. The recovery phase of the living organism after surgical intervention takes into account the long-term resistance of the implant [13, 32].

Regardless of the medical application, making a biomaterial depends on several requirements: it is not toxic, it does not contain filter products, it does not cause allergic, carcinogenic, teratogenic effects (generated by morphological abnormalities), it does not cause rejection phenomena body, do not alter the blood composition and not disturb the coagulation mechanism, do not modify the biological pH, do not contain hydrophobic or hydrophobic sites that promote cellular penetration and adhesion [3, 11, 22, 26].

II.3. Titanium and Ti-based Alloys

Titanium is a gray metallic element having the atomic number Z = 22 in the periodic element system, being in the fourth group IV (Figure II.1).

The pure titanium has a melting point of 1668°C and a boiling point of 3287°C. Because of the relatively high temperature of the melting point it also makes it useful as a refractory metal.

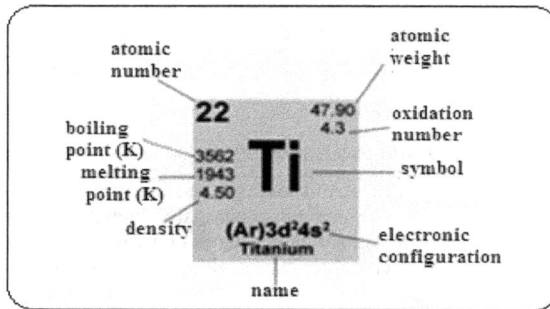

Figure II.1. Titanium [33].

The physical properties of titanium are particularly noticeable by the favorable ratio between mechanical strength and density, high melting point, low thermal conductivity, and high surface tension in the melted state, with major influences on hot processing of the material. The main physical properties of pure titanium are shown in Table II.5.

The titanium is relatively rough, paramagnetic and a weak heat and electricity leader.

The mechanical properties of titanium are determined by its degree of purity, which fundamentally depends on the technological conditions of production and processing. The presence of impurities in determined amounts increases hardness and resistance to the detriment of plasticity. The presence of impurities is also determined by the increased affinity of titanium for oxygen, nitrogen, carbon and hydrogen at high temperatures [25, 34-36].

Table II.5. The physical properties of titanium [36].

Property	Characteristic/ Value
Color in compact condition	Silver - gray
Density at 25°C (α-Ti)	4.51g/cm3
Density at 900°C (β-Ti)	4.33g/cm3
Melting temperature	1677°C
Coefficient of thermal expansion	9.1x10-6/K
Specific heat at 25°C	0,523 J/g.K
Thermal conductivity at 25°C	17-22 W/mK
Surface tension at 1600°C	1.7 N/m
Elasticity at 25°C	108 GN/m2
Tensile strength resistant	450 MPa before casting, respectively 850 MPa, after casting
Stretching limit	100-200 N/m2; 15-20%
Hardness	160-190 HB, 80-105 HV

Titanium used in medicine has four degrees of purity, differentiated by number and concentration of impurities. Purity grades of "pure commercial" titanium are shown in Table II.6.

Table II.6. Purity grades of "pure commercial" titanium [35].

Type	N_{max}	Fe_{max}	O_{max}	C_{max}	H_{max}	Ti
Grade 1	0.03	0.20	0.18	0.10	0.015	balance
Grade 2	0.03	0.30	0.25	0.10	0.015	balance
Grade 3	0.05	0.30	0.35	0.10	0.015	balance
Grade 4	0.05	0.50	0.40	0.10	0.015	balance

Titanium with purity grade 1 and 2 is used in dentistry for fixed prostheses, while titanium grade 4 is used to make skeleton prostheses. For the extensive use of titanium as a biomaterial in the human body it is recommended to add biocompatible elements such as aluminum, vanadium, zirconium, tantalum, niobium, etc. [34, 35].

Processing of titanium by melting / casting can expose the metal mass to different amounts of impurities, which can influence its mechanical properties. The mechanical properties of titanium of different grades are highlighted in Table II.7.

Table II.7. The mechanical properties of titanium of various grades [25].

Type Ti	Tensile Strength, (MPa)	Flow limit, (MPa)	Hardness Vickers, (HV)
Ti grade 1	290-410	180	126
Ti grade 2	390-540	250	158
Ti grade 3	460-590	320	179
Ti grade 4	540-740	390	211

Titanium chemical properties represent a different interest as it explains the technological particularities of the material and its special behavior in biological environments. Titanium is a very reactive metal (immediately after aluminum in the metal reactivity series). Also, it does react intensively in contact with gases, especially at high temperatures [33].

II.4. Titanium Used in Medical Applications

The use of biomaterials in medical applications comprises several areas: orthopedics, cardiovascular surgery, ophthalmology, dentistry, urology, aesthetic surgery, neurology, suture material for wound healing, controlled drug delivery systems [4, 19, 37-41].

Biomedical implants are supposed to have various requirements demanded by human body, because they could be applied to almost any part of the body. Biomechanical properties like stiffness, strength, fracture toughness, wear resistance, fatigue strength, corrosion resistance and biomedical properties like toxicity, surface state, osseointegration are important characteristics that an implant requires [37-41].

In the past few decades, titanium and its alloys have received great attention as biomedical materials, used in load bearing implants such as artificial hip joints, bone plates and screws, spinal instruments and dental implants [3]. The main advantages of titanium based biomaterials are the following: acceptable biocompatibility, low relative longitudinal elastic modulus, excellent corrosion resistance because of oxide's coating, a very good density/strength ratio [41].

These alloys are used as implants such as dental implants and roots (Figure II.2a), devices replacing failed hard tissue, artificial hip joints (Figure II.2b), bone plates (Figure II.2c), screws for fracture fixation (Figure II.2d), cardiac valve prostheses (Figure II.2 e), pacemakers and artificial hearts.

Most researchers have reached the conclusion that the most important factors for a biomaterial are: mechanical properties, biocompatibility and corrosion properties. Patient health and surgery competence also contribute to the success of the biomaterial [11, 41].

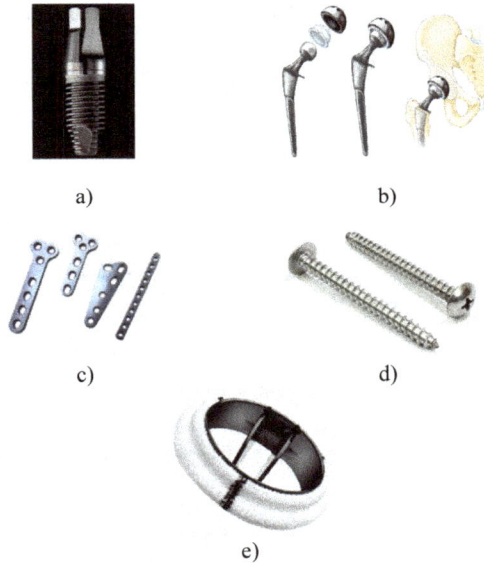

a) b)

c) d)

e)

Figure II.2. Examples of medical applications of Ti-based alloys: a) dental implants, b) artificial hip joint, c) bone plates, d) screws, e) cardiac valve prostheses [36-41].

II.5. Classification of Titanium Alloys and Their Properties

Titanium is an allotropic material, exhibiting different forms and properties: up to a temperature of 882°C, having a hexagonal-compact structure (α) and above 882°C having a centered cube structure (β) [42].

The influence of alloying elements on titanium alloys contributes to a wide range of different microstructural and mechanical properties. Thus, the alloying elements are divided into three categories:

- α stabilizing: C, N_2, O_2, Al;
- β stabilizing: V, Nb, Mo, Ta, Fe, Mn, Cr, Co, W, Ni, Cu, Si, H_2;
- Neutral elements: Zr, Sn, Hf, Ge, Th.

As the alloying elements are added to the titanium, a change occurs in the temperature at which the phase transformation takes place. Alloying the pure titanium with α stabilizing elements increases the temperature range in which the α phase exists, respectively the alloy with elements β leads to the increase of the β phase domain. Other elements such as Zr or Sn have a neutral contribution to the temperature domains in which the two phases exist [6, 19, 43-45].

As a result, titanium alloys are divided into three classes, depending on the structure and alloying elements (Figure II.3).

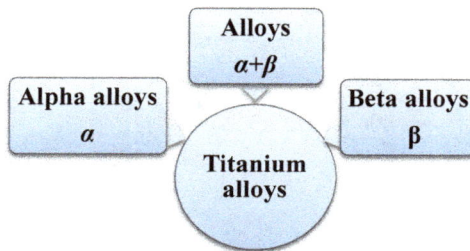

Figure II.3. Classification of titanium alloys [43-45].

Alpha alloys are alloys with aluminum, oxygen and / or nitrogen addition, which generally stabilize the "alpha" phase (C.P.-Ti, Ti5Al2.5Sn) [19, 24]. These, in general, cannot be heat treated, but can be welded easily. It has a medium resistance, acceptable ductility, resilience and good mechanical properties.

The α + β alloys have a structure consisting of the two alpha and beta phases (Ti6Al4V, Ti6Al7Nb) [19, 24]. These alloys can be heat-treated and welded, but the high-temperature creep limit is not as good as most alpha alloys.

Beta alloys are alloys with addition of molybdenum, iron, vanadium, chromium and / or manganese, which generally stabilizes the beta phase (Ti13Nb13Zr, Ti15Mo) [16, 21]. These alloys can be heat treated and welded easily.

The study of titanium alloys has been and is an intense concern of researchers with the aim of improving them, of the mechanical, chemical and biological characteristics imposed on implant materials. Titanium alloys were developed by alloying with non-toxic elements to the human body, replacing aluminum and vanadium with other biocompatible elements with human tissue such as molybdenum, zirconium, tantalum, niobium and silicon. Thus, beta alloys and α + β alloys resulted in superior biocompatibility and improved mechanical properties [6, 7, 46-49].

II.5.1 Microstructure of Ti-Mo alloys

Microstructure and mechanical properties of Ti-Mo alloys differ depending on the amount of stabilizing elements α or β added in the chemical composition.

Figure II.4 illustrates the microstructure of Ti-Mo alloys of beta type. Zhou et al. [50] developed two Ti10Mo and Ti20Mo alloys, highlighting the formation of beta-equiaxial grains of different sizes, highlighting that the 10% Mo alloy has grains higher than the alloy with a concentration of 20% Mo.

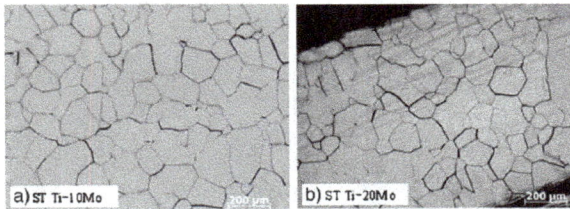

Figure II.4. Microstructure of beta-type alloys: a) Ti10Mo, b) Ti20Mo [50].

Gordin and collaborators [51] studied the influence of alloying elements, tantalum and molybdenum on the Ti12Mo5Ta alloy by comparing its microstructure with that of the Ti6Al4V alloy (Figure II.5).

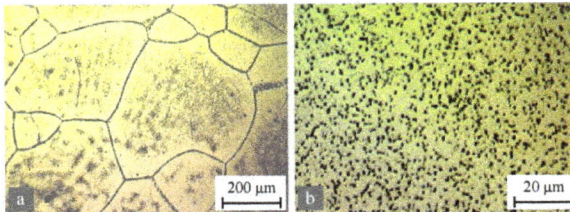

Figure II.5. Optical micrographs of alloys: a) Ti12Mo5Ta, b) Ti6Al4V [51].

From the optical micrographs of these alloys, the beta-cubic with centered volume microstructure of the Ti12Mo5Ta alloy and an alpha + beta equiaxial structure attributed to the Ti6Al4V alloy were observed. The presence of aluminum contributed to the stabilization of the alpha phase at low temperatures, and vanadium to the stabilization of the beta phase. The beta-cubic microstructure of the Ti12Mo5Ta alloy is influenced by the presence of the two molybdenum and tantalum stabilizing beta elements.

Zhan et al. [52] highlighted the microstructure of two beta Ti10Mo1.25Si4Zr and Ti10Mo1.25Si7Zr alloys (Figure II.6), consistent with the solubility of alloying elements, and it was observed that the alloy microstructure contained beta grains. Alloying with zirconium and silicon led to the limit of the temperature range of the alpha phase, keeping the beta phase in Ti10Mo alloys down to low temperatures. It has also been found that increasing the zirconium concentration leads to refining the microstructure, the grain size for Zr = 4% being about 160 microns and 100 microns for Zr = 7%.

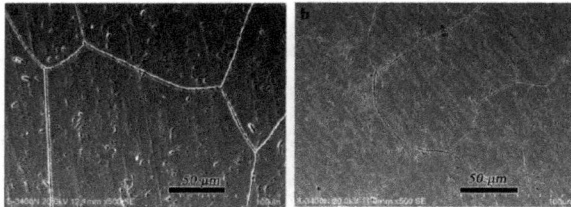

Figure II.6. SEM images of beta alloys: a) Ti10Mo1.25Si4Zr, b) Ti10Mo1.25Si7Zr [52].

It is important to know the microstructure of titanium biomaterials because the coexistence of the two phases is an important factor that controls mechanical properties and corrosion resistance, having a direct effect on biocompatibility [19, 53-55].

II.5.2 Mechanical Properties of Ti-Mo Alloys

Alloying elements contribute to changing the characteristics of titanium alloys, primarily by modifying the mechanical properties and also the cytotoxicity.

The first generation of titanium based biomaterials, developed before 1990 (C.P. Ti, Ti6Al4V, Ti6Al7Nb), has shown numerous allergic reactions in the human body induced by aluminum and vanadium elements, highlighted by many studies [13, 14].

Song and collaborators [56] pointed out that niobium, zirconium, molybdenum and tantalum are best suited as alloying elements in order to reduce the modulus of elasticity of cubic titanium without altering mechanical strength. Also, an important aspect is that these alloying elements fall into the category of non-toxic elements, giving them the advantage of being used for applications in implantology [57]. Based on these issues, researchers have developed and researched the second generation of biomaterials [15, 53], focusing on the development of biocompatible titanium alloys, improved alloys with non-toxic elements to the human body (Mo, Zr, Ta) by developing materials such as metastable beta alloys: Ti15Mo, Ti15Mo5Zr3Al, Ti10Mo1.2Si13Zr, etc.

Table II.8 shows the main mechanical properties for several Ti-Mo alloys compared to classical biomaterials (stainless steel, CoCr alloys).

Table II.8. The mechanical properties of Ti-Mo biomaterials [33, 41].

Material (% mass)	Tensile Strength (MPa)	Flow limit (MPa)	Elongation (%)	The modulus of elasticity (GPa)	Type of alloy
First generation of biomaterials (1950-1990)					
C.P. Ti (C.P. grad 1-4)	240	170	24	102.7	α
Ti6Al4V	860-930	825-869	6-10	112	α + β
Ti6Al7Nb	860	795	10	105	α + β
Ti5Al2.5Fe	690	585	15	100	α + β
The second generation of biomaterials (1990 to today)					
Ti-12Mo-5Zr-2Fe	1060-1100	1000-1060	18-22	74-85	β
Ti15Mo	874	544	21	78	β
Ti15Mo5Zr3Al	852-1100	838-1060	18-25	80	β
Ti10Mo1.2Si4Zr	1412	825	39	23	β
Ti10Mo1.2Si13Zr	1417	1024	29	32	β
Ti10Mo	1092	507	36	25	β
Ti10Mo1.7Si	1205	712	21	21	β
Ti12Mo5Ta	-	-	-	74	β
Ti10Mo3Nb	1918	1854	10	28	β
Ti10Mo10Nb	1717	1404	30	24	β
316L	500-1350	200-700	10-40	200	-
CoCr Alloys	900-1000	900-1800	500-1500	240	-
Human Bone	100-300	-	1-2	30	-

Ti-Mo alloys alloyed with various biocompatible elements compared to other classical biomaterials, have superior mechanical properties such as high tensile strength and a much lower modulus of elasticity close to that of human bone.

According to the data presented in Table II.8, it can be noticed that after the 1990s, Ti-Mo alloys were studied and improved with different alloying elements. The authors correlated the values of the mechanical characteristics with the X-ray diffraction results and pointed out that the low elastic modulus alloys are due to the presence of the beta phase [19].

By adding different alloying elements, different mechanical properties resulted, as outlined in Table II.9.

The values of the longitudinal elasticity modulus of the Ti-Mo alloys compared to the classical ones are shown in Figure II.7. β -type alloys have much lower longitudinal elastic modulus compared to α or α + β alloys, which is why their research and development is a priority for researchers.

Table II.9. Influence of alloying elements on mechanical properties [13, 19].

Alpha alloys	Beta alloys
Low density	High density
Low resistance	High resistance
High elasticity	Low modulus of elasticity

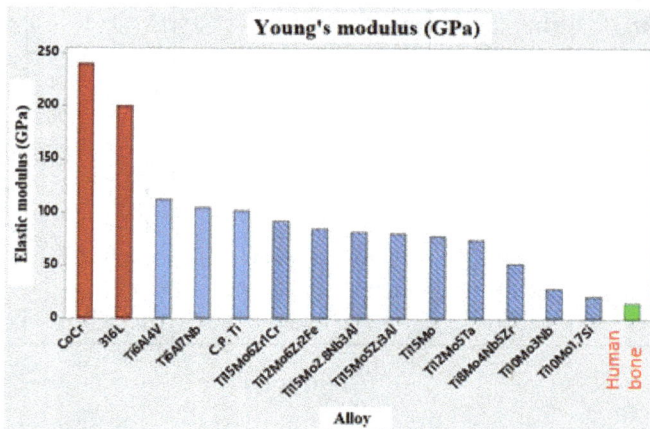

Figure II.7. Values of the longitudinal elastic modulus for titanium alloys [58].

Studies [19, 24, 56] have shown that α + β alloys possess satisfactory characteristics, in particular superior mechanical properties, as well as increased alloy α resistance to both corrosion and oxidation. On the other hand, β -alloys, due to molybdenum stabilizing elements, tantalum and niobium, have the advantage of increasing mechanical strength and an elastic modulus close to that of the human bone, important aspects of long-term use of biomaterials in the medical field.

II.5.3 Corrosion of Ti-Mo Alloys

Corrosion resistance is an essential feature of titanium and its alloys due to its thin, compact and extremely stable titanium dioxide film that forms in few seconds after contact with the environment. Thus, due to the very good corrosion resistance, titanium is used in medicine, resists water, acids or salt solutions, having a similar behavior to platinum in case of chemical corrosion [11, 59].

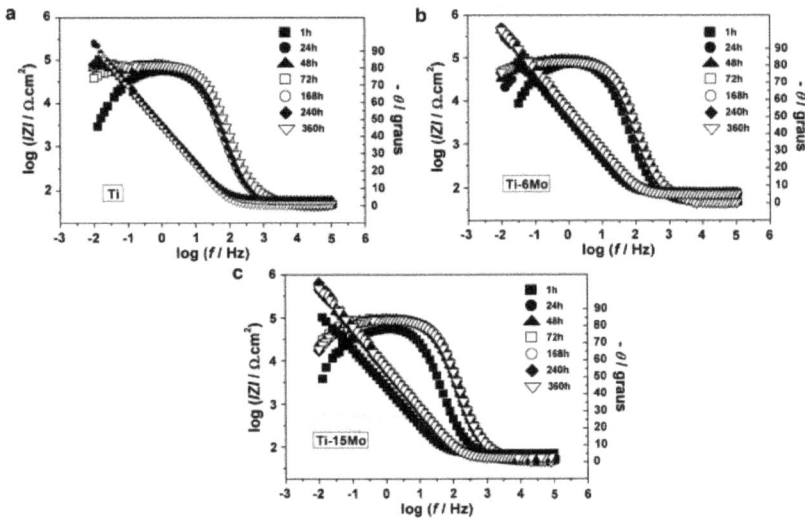

Figure II.8. Bode charts for: a) C.P. Ti, b) Ti-6Mo; c) Ti-15Mo at different immersion time ranges in Ringer's solution [60].

Oliveira et al. [60] subjected for electrochemical testing in Ringer's solution, three types of biocompatible alloys: C.P. Ti, Ti6Mo and Ti15Mo at different time intervals (Figure II.8). Electrochemical behavior has shown that there are no major changes on the surface

19

of samples immersed 24 hours in the solution used, forming titanium oxide layers of different thicknesses. The authors argued that alloying up to 15% molybdenum leads to improved corrosion protection characteristics and the concentration above 20% leads to the opposite effect.

Another comparative study between Ti6Al7Nb and Ti30Ta, Ti40Ta, Ti50Ta, Ti60Ta was performed [59, 61] in various types of saliva (artificial saliva, acidified saliva, fluorinated saliva). The polarization curves in the different environments are shown in Figure II.9.

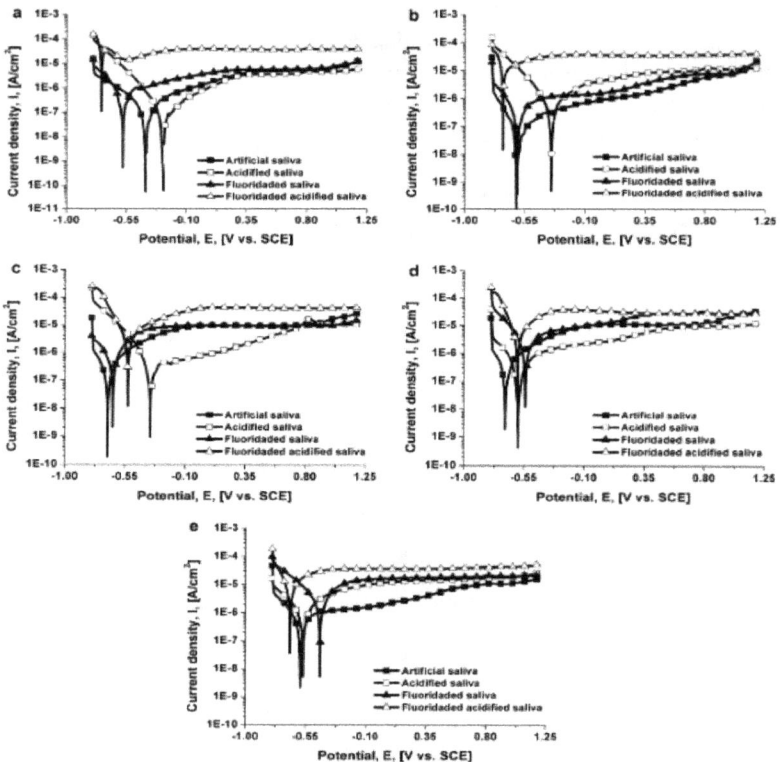

Figure II.9. Polarization curves in different environments: a) Ti6Al7Nb; b) Ti30Ta; c) Ti40Ta; d) Ti50Ta; e) Ti60Ta [61].

The Ti6Al7Nb alloy exhibited a lower corrosion resistance and a less stable oxide layer for the fluorinated saliva. TiTa alloys have shown high corrosion resistance due to the formation of Ta_2O_5 oxide on the alloy surface, which has contributed to increased corrosion protection.

Passivation of these alloys is influenced by the presence of alloying elements as well as the presence of impurities in the material. Although titanium, in the context of the electrochemical series, is a metal that should corrode strongly, but due to the formation of a oxide layer, it remains passive in the human body, which makes it very useful in medical applications [61, 62].

II.5.4 Biocompatibility of Ti-Mo Alloys

In the human body, titanium is found on average in an amount of 10-20 mg, with the highest titanium concentration being in pulmonary tissue. This concentration was found to be increased six-fold in the case of implantation of titanium rods and also, in the use of osteosynthesis devices. Daily titanium ingestion varies from 0.1 mg to 1 mg depending on the amount of drinking water and food used in alimentation [24, 29].

José Roberto Severino Martins Júnior et al. [63] studied the Ti15Mo alloy and the commercially pure titanium alloy, performing a cytotoxicity assay. Figure II.10 highlights the test results, indicating that these alloys are not highly toxic and biocompatible with human tissue. The authors also studied cell morphology, indicating that the material did not cause any aggression, and also that no change in cell morphology or adhesion to the surface of the material was induced.

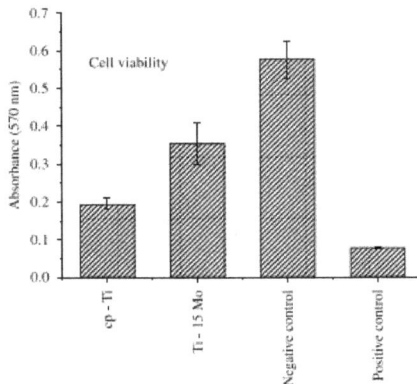

Figure II.10. Cell viability for the Ti15Mo alloy, compared to an alloy of C.P. Ti [63].

The Ti-Zr binary alloy [64], already marketed and used in the field of implantology, has been confirmed through studies to induce a low inflammatory response at the subcutaneous interface compared to the pure titanium alloy, thus demonstrating superior biocompatibility.

Correea et al. [64] performed cell viability cytotoxicity assays (Figure II.11) on Ti5Zr, Ti10Zr and Ti15Zr alloys, and cell viability was found to be relatively similar for the three samples compared to pure titanium, as shown in Figure I.15. Also, the use of zirconium as an alloying element also contributes significantly to the increase in mechanical properties.

The biocompatibility of titanium and its alloys is closely related to the formation of oxide layers (TiO_2) on the surface of the implant [65], as an interface between the implant and the biological environment. Alloy elements such as Zr, Ta, Nb, Sn do not affect cell viability and have a reduced amount of ions released in the body, but Al and V contribute to reduce cell viability [65]. Other elements such as Ag, Co, Cr, Cu have moderate cytotoxicity behavior, but their presence in these alloys significantly reduces their toxicity [66, 67].

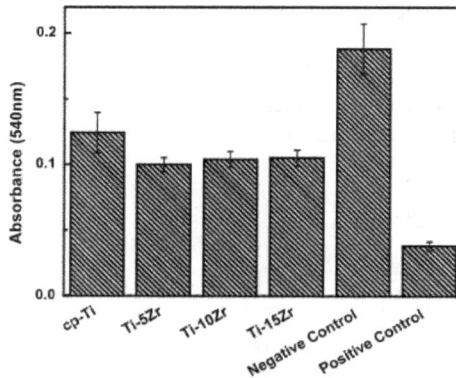

Figure II.11. The results of cytotoxicity tests of Ti-Zr alloys compared to C.P. Ti [64].

In the field of medicine, a wide variety of biomaterials are used in the form of medical devices, implants and prosthesis systems. If it is relatively easy to find a material that meets certain functional requirements, it is very difficult to find one that is able to maintain its performance for a long time without deterioration, with unwanted effects induced in the body and properties close to the system it replaces.

When developing an alloy for medical applications, the main features such as biocompatibility, mechanical properties and corrosion resistance must be considered. Uses of titanium in medical applications were due to superior mechanical properties of classical alloys (stainless steel, CoCr alloys): low elasticity, corrosion resistance due to the formation of a protective oxide layer and its good tolerance in the human body.

There are three types of titanium based alloys: α, α + β and β. Alloys α and α + β alloys have improved characteristics for mechanical properties, increased corrosion resistance and oxidation. The β-type alloys, due to the Mo, Ta and Nb stabilizing elements, have the advantage of increasing mechanical strength and a modulus of elasticity close to that of the human bone.

Current researchers' analysis provided information on all alloying elements used for titanium alloys with possible medical applications. The alloying elements have been studied in terms of influence on mechanical and chemical properties. Thus, in order to develop a titanium alloy, elements exhibiting high biocompatibility with human tissue were selected and improved mechanical properties: Mo, Ta and Zr. Studies have shown that molybdenum and tantalum help reduce the elastic modulus and show high corrosion resistance, and zirconium helps to refine the structure.

The development of new titanium alloy variants opens new applicative opportunities, with the possibility of patenting chemical compositions and technological transfer.

References

[1]   N. Dumitraşcu, Biomateriale şi biocompatibilitate, Ed. Universităţii "Alexandru Ioan Cuza", Iaşi, 2007.

[2]   D.P. Balaban, Biomateriale, Ed. Ovidius University Press, Constanţa, 2005.

[3]   C. Mihăilescu, A. Mihăilescu, N. Georgescu, Biomateriale de uz chirurgical, Ed. Tehnopress, Iaşi, 2011.

[4]   C.N. Cumpătă, Implaturi acoperite chimic cu hidroxiapatita biologică, Ed. Printech, Bucureşti, 2012.

[5]   I. Bălţatu, P. Vizureanu, F. Ciolacu, D.C. Achiţei, M.S. Bălţatu, D. Vlad, In Vitro study for new Ti-Mo-Zr-Ta alloys for medical use, IOP Conf. Ser.: Mater. Sci. Eng., 572 (2019) 012030. https://doi.org/10.1088/1757-899X/572/1/012030

[6]   M.S. Baltatu, P. Vizureanu, T. Balan, M. Lohan, C.A. Tugui, Preliminary Tests for Ti-Mo-Zr-Ta Alloys as Potential Biomaterials, Book Series: IOP Conference Series-

Materials Science and Engineering, 374 (2018), 012023. https://doi.org/10.1088/1757-899X/374/1/012023

[7]   M.S. Bălțatu, P. Vizureanu, V. Goanță, C.A. Tugui, I. Voiculescu, Mechanical tests for Ti-based alloys as new medical materials, IOP Conf. Ser.: Mater. Sci. Eng., 572 (2019) 012029. https://doi.org/10.1088/1757-899X/572/1/012029

[8]   C. Lipșa, D.S. Lipșa, Biomateriale-Curs pentruanul I, Iași, 2009.

[9]   D.C. Ludwigson, Requirements for metallic surgical implants and prosthetic devices. Metals Engineering Quarterly: American Society of Metallurgists 1, 1965.

[10]   C. Demian, Cercetări privind comportarea materialelor destinate implantării osoase conform normelor europene de calitate - teză de doctorat, Universitatea "Politehnica", Timișoara, 2007.

[11]   R. Chelariu, G. Bujoreanu, C. Roman, Materiale metalice biocompatibile cu baza titan, Ed. Politehnium, Iași, 2006.

[12]   I. Antoniac, Biomateriale metalice utilizate la executia componentelor endoprotezelor totale de sold, Ed. Printech, București, 2007.

[13]   D.M. Bombac, M. Brojan, P. Fajfar, F. Kosel, R. Turk, Review of materials in medical applications, Materials and Geoenvironment, 54(4), 2007, pp. 471-499.

[14]   C.N. Elias, J.H.C. Lima, R. Valiev, M.A. Meyers, Biomedical applications of titanium and its alloys, Biological Materials Science, 2008, pp. 46- 49. https://doi.org/10.1007/s11837-008-0031-1

[15]   M. Niinomi, Mechanical properties of biomedical titanium alloys, Mater. Sci. Eng., A, 243 (1998) 231-236. https://doi.org/10.1016/S0921-5093(97)00806-X

[16]   L.I. Linkow, Prefabicated mandibular prostheses for intraosseous implants, J Prosthet Dent., 20(4), 1968, pp. 367–375. https://doi.org/10.1016/0022-3913(68)90234-5

[17]   L.I. Linkow, The blade vent-a new dimension in endosseous implantology, Dent Concepts, 11 (2), 1968, pp.3-12.

[18]   G. Lűtjering, J.C. Williams, Titanium-Second Edition, Springer Science + Business Media, Germany, 2000.

[19]   M. Geetha M., A.K. Singh, R. Asokamani, A.K. Gogia, Ti based biomaterials, the ultimate choice for orthopaedic implants - A review, Mater. Sci., 54 (2009) 397-425. https://doi.org/10.1016/j.pmatsci.2008.06.004

[20]   D. Bunea, A. Nocivin, Materiale biocompatibile, Ed. și Atelierele Tipografice Bren, București, 1998.

[21]   H. Vermeșan, Cercetări privind comportarea la coroziune a otelurilor inoxidabile

supuse deformarii plastice si nitrurarii ionice - teza de doctorat, Universitatea Tehnică din Cluj Napoca, 1998.

[22] C. Popa, V. Cândea, V. Șimon, D. Lucaciu, O. Rotaru, Știința biomaterialelor, Ed. U.T. Press, Cluj-Napoca, 2008.

[23] M.G. Minciună, Contribuții privind îmbunătățirea proprietăților aliajelor de cobalt utilizate în aplicații medicale-Teză de doctorat, Iași, 2014.

[24] Q. Chen, G.A. Thouas, Metallic implant biomaterials, Materials Science and Engineering R, 87 (2015) 1–57. https://doi.org/10.1016/j.mser.2014.10.001

[25] C. Leyens, M. Peters, Titanium and Titanium alloys. Fundamentals and Applications, Ed. Wiley-VCH, 2003, pp. 423-431. https://doi.org/10.1002/3527602119

[26] M.S. Bălțatu, P. Vizureanu, V. Geantă, C. Nejneru, C.A. Țugui, S.C. Focșăneanu, Obtaining and Mechanical Properties of Ti-Mo-Zr-Ta Alloys, IOP Conference Series: Materials Science and Engineering, 209 (2017) 012019. https://doi.org/10.1088/1757-899X/209/1/012019

[27] F. Miculescu, Tehnici de analiză și control a biomaterialelor, Ed. Printech, București, 2009.

[28] A.C. Bărbînță, Îmbunătățirea proprietăților aliajelor de Ti-Nb-Zr-Ta utilizate la fabricarea protezelor ortopedice - teză de doctorat, Iași, 2003.

[29] R.A. Roșu, Metode de obținere și de prelucrarea biomaterialelor pentru proteze umane - teză de doctorat, Universitatea "Politehnica", Timișoara, 2008.

[30] J.R. Davis, Handbook of materials for medical devices, ASM International, United States of America, 2003.

[31] C.N. Cumpătă, Implaturi acoperite chimic cu hidroxiapatita biologică, Ed. Printech, București, 2012.

[32] E. Pincovschi, C.M. Florea, Compuși anorganici biocompatibili cu aplicații în implantologie, Ed. Printech, București, 1997.

[33] M.S. Cercel (cas. Bălțatu), Contribuții privind îmbunătățirea proprietăților aliajelor de Ti-Mo destinate aplicațiilor medicale – teză de doctorat, Iași, 2017.

[34] S.U. Țuculescu, E. Bratu, S. Lakatos, Titanul în stomatologie, Ed. Signata, Timișoara, 2001.

[35] M. Dobrescu, C. Dumitrescu, M. Vasilescu, Titan și aliaje de titan, Ed. Printech, București, 2000.

[36] K. Wang, The use of titanium for medical applications in the USA, Mater. Sci. Eng. A, 223 (1996) 134-137. https://doi.org/10.1016/0921-5093(96)10243-4

[37] Information on https://orthoinfo.aaos.org

[38] Information on http://n2-uk.com

[39] Information on http://Amazon.com

[40] Information on http://indiamart.com

[41] M.S. Bălțatu, P. Vizureanu, M.H. Țierean, M.G. Minciună, D.C. Achiței, Ti-Mo Alloys used in medical applications, Advanced Materials Research, 1128 (2015)105-111. https://doi.org/10.4028/www.scientific.net/AMR.1128.105

[42] ***ASM Handbook, Alloy Phase Diagrams, vol.3.

[43] E.W. Collings, The Physical Metallurgy of Titanium Alloys, American Society for Metals, 1984.

[44] M.J. Donachi, Titanium: A Technical Guide, ASM INTERNATIONAL, 1988.

[45] *** ASM Handbook, Properties and Selected Nonfferous Alloys and Special-Purpose Materials, Vol. 2.

[46] S.G. Steinemann, P.A. Mausli, S. Szmukler-Moncler, M. Semlitsch, O. Pohler, H.E. Hintermann, et al. Beta titanium in the 1990s.Warrendale, Pennsylvania: The Mineral, Metals and Materials Society, 1993, pp. 2689–1696.

[47] Y.J. Park, Y.H. Song, J.H. An, H.J. Song, K.J. Anusavice, Cytocompatibility of pure metals and experimental binary titanium alloys for implant materials, Journal of dentistry, 41 (2013) 1251-1258. https://doi.org/10.1016/j.jdent.2013.09.003

[48] L.B. Zhang, K.Z. Wang, L.J. Xu, S.L. Xiao, Y.Y. Yu-yong Chen., Effect of Nb addition on microstructure, mechanical properties and castability of type TiMo alloys, Trans Nonferrous Met. Soc. China, 25 (2015) 2214-2220. https://doi.org/10.1016/S1003-6326(15)63834-1

[49] D.R.N. Correa, F.B. Vicente, A.R. Oliveira, M.L. Lourenc, P.A.B. Kuroda, M.A.R. Buzala, C.R. Grandini, Effect of the substitutional elements on the microstructure of the Ti-15Mo-Zr and Ti-15Zr-Mo systems alloys, J. Mater Res Technol, vol. 4(2), 2015, pp.180-185. https://doi.org/10.1016/j.jmrt.2015.02.007

[50] Y.L. Zhou, D.M. Luo D.M., Microstructures and mechanical properties of Ti–Mo alloys cold-rolled and heat treated, Materials characterization 62 (2011) 931-937. https://doi.org/10.1016/j.matchar.2011.07.010

[51] D.M. Gordin, T. Gloriant, G. Nemtoi, R. Chelariu, N. Aelenei, A. Guillou, D. Ansel, Synthesis, structure and electrochemical behavior of a beta Ti-12Mo-5Ta alloy as new biomaterial, Materials Letters, 59 (2005) 2936 – 2941. https://doi.org/10.1016/j.matlet.2004.09.063

[52] Y. Zhan, C. Li, W. Jiang, β-type Ti-10Mo-1.25Si-xZr biomaterials for applications in hard tissue replacements, Materials Science and Engineering C, 32 (2012) 1664–1668. https://doi.org/10.1016/j.msec.2012.04.059

[53] Bălțatu M.S., Vizureanu P., Istrate B., Physical and structural characterization of Ti-based alloy, International Journal of Modern Manufacturing Technologies, (2015), Vol. VII, No. 2, pp.12-17.

[54] M.S. Bălțatu, R. Cimpoeşu, P. Vizureanu, D.C. Achitei, M.G. Minciuna, Microstructural characterization of TiMoZrTa alloy, The Annals of „Dunărea de Jos" University of Galaţi, Fascicle IX. Metallurgy and materials science, 4 (2015) 23-26.

[55] M.S. Bălțatu, P. Vizureanu, M. Benchea, M.G. Minciună, A.C. Achiței, B. Istrate, Ti-Mo-Zr-Ta Alloy for Biomedical Applications: Microstructures and Mechanical Properties, Key Engineering Materials, 750 (2018)184-188. https://doi.org/10.4028/www.scientific.net/KEM.750.184

[56] Y. Song, D.S. Xu, R. Yang, D. Li, W.T. Wu, Z.X. Guo, et al. Mater Sci Eng A, vol. 260 (1999) 269–74. https://doi.org/10.1016/S0921-5093(98)00886-7

[57] S.J. Li, R. Yang, S. Li, Y.L. Hao, Y.Y. Cui, M. Niinomi, et al. Wear, 257 (2004) 869–876. https://doi.org/10.1016/j.wear.2004.04.001

[58] M.S. Baltatu, C.A. Tugui, M.C. Perju, M. Benchea, M.C. Spataru, A.V. Sandu, P. Vizureanu, Biocompatible Titanium Alloys used in Medical Applications, Revista de Chimie, Bucharest, 70(4), 2019, pp. 1302-1306. https://doi.org/10.37358/RC.19.4.7114

[59] M.S. Bălțatu, P. Vizureanu, D. Mareci, L.C. Burtan, C. Chiruţă, L.C Trincă, Effect of Ta on the electrochemical behavior of new TiMoZrTa alloys in artificial physiological solution simulating in vitro inflammatory conditions, Materials and Corrosion, vol. 67(12), 2016, pp. 1314-1320. https://doi.org/10.1002/maco.201609041

[60] N.T.C. Oliveira, A.C. Guastaldi, Electrochemical stability and corrosion resistance of Ti–Mo alloys for biomedical applications, Acta Biomaterialia, 5 (2009) 399–405. https://doi.org/10.1016/j.actbio.2008.07.010

[61] D. Mareci, R. Chelariu, D.M. Gordin, G. Ungureanu, T. Gloriant, Comparative corrosion study of Ti–Ta alloys for dental applications, Acta Biomaterialia, 5 (2009) 3625–3639. https://doi.org/10.1016/j.actbio.2009.05.037

[62] M.S. Bălțatu, P. Vizureanu, R. Cimpoeşu, M.M.A.B. Abdullah, A.V. Sandu, The Corrosion Behavior of TiMoZrTa Alloys Used for Medical Applications, Revista de Chimie, 67 (10), 2016, pp. 2100-2002.

[63] J.R.S.M. Junior, R.A. Nogueira, R. Oliveira de Araújo, T.A.G. Donato, V.E.A. Chavez, A.P.R.A. Claro, J.C.S.M. Moraes, M.A.R. Buzalaf, C.R. Grandini,

Preparation and Characterization of Ti-15Mo Alloy used as Biomaterial, Materials Research, vol.14(1), 2011, pp.107-112. https://doi.org/10.1590/S1516-14392011005000013

[64] D.R.N. Correa, F.B. Vicente, T.A.G. Donato, V.E. Arana-Chavez, M.A.R. Buzalaf, C.R. Grandini, The effect of the solute on the structure, selected mechanical properties,and biocompatibility of Ti–Zr system alloys for dental applications, Materials Science and Engineering C, 34 (2014) 354–359. https://doi.org/10.1016/j.msec.2013.09.032

[65] Y.H. Jeong, H.C. Choe, W.A. Brantley, Nanostructured thin film formation on femtosecond laser-textured Ti-35Nb-xZr alloy for biomedical applications, Thin Solid Films, vol. 519 (15), 2011, pp. 4668–4675. https://doi.org/10.1016/j.tsf.2011.01.014

[66] Y. Okazaki, S. Rao, Y. Ito, T. Tateishi, Corrosion resistance, mechanical properties: corrosion fatigue strength and cytocompatibility of new Ti alloys without Al and V, Biomaterials, vol.13 (19), 1998, pp. 1197–1215. https://doi.org/10.1016/S0142-9612(97)00235-4

[67] Y.H. Song, M.K. Kim, E.J. Park, H.J. Song, K.J. Anusavice, Y.J. Park, Cytotoxicity of alloying elements and experimental titanium alloys by WST-1 and agar overlay tests, Dent. Mater., vol. 30 (9), 2014, pp. 977–983. https://doi.org/10.1016/j.dental.2014.05.012

CHAPTER III: Objectives and Methodology of Experimental Research

Over the years, some materials have been developed and researched for making implants for medical applications, of which few have been accepted by the human body, namely those that have certain properties needed to achieve long-term success. Each biomaterial or device must meet mechanical and performance requirements, which originate from the need to perform physiological function that is appropriate to the properties of the material [1-4].

The final material must fulfill certain technological and economic requirements, which can be evaluated mainly by the analysis of the following factors: composition, structure, properties and performances [5-7].

III.1. Objectives

Improving the properties of biomaterials is a necessity to reduce the failure rate of implants in human tissue as a result of low cell adhesion and inappropriate mechanical properties.

Titanium and its alloys have lately become an important alternative [5-12] in medical applications being considered the safest material for implantation. An increased interest in their application in the medical field has been observed taking into account the properties they present (mechanical properties, low density, good biocompatibility, increased corrosion resistance, etc.).

The main objective of this book is to obtain and characterize new alloys with different concentrations of non-toxic chemical elements (Ti, Mo, Zr, Ta). In line with this objective, a number of activities have resulted:

(i) the design and system choice with elements like Ti, Mo, Zr and Ta;

(ii) study of the influence of the alloying elements in the process of obtaining the alloys proposed and the establishment of the elaboration's method;

(iii) obtaining;

(iv) characterization of new materials through laboratory tests to identify all their properties.

Considering the materials / alloys used in medical applications, six original alloy variants have been designed / developed to achieve the goal of this book in order to improve the mechanical properties (mechanical strength, modulus of elasticity, hardness, compressive strength), corrosion resistance and showing appropriate cytotoxicity tests for human tissue. The composition of alloys based on titanium, molybdenum, zirconium and tantalum was designed according to the influence of these elements on alloy characteristics, biocompatibility with the human body, and influence on corrosion behavior in biological fluid.

In order to achieve the main goal of the book, a research program described in Table III.1 was developed and implemented which aims at elaboration and characterization of TMZT alloys (TiMoZrTa).

Table III.1. The experimental research program [5].

1. Current state of the art
Considering an extensive bibliographic material on current biomaterial research has allowed the study of all classes of biomaterials together with their properties for medical applications. Critical analysis of the current state of research has led to the study of Ti-Mo alloys with possible medical applications to overcome the disadvantages of existing metallic biomaterials (high longitudinal elasticity, low corrosion resistance, low biocompatibility).
2. Design of TMZT alloys
Design and selection of new advanced materials based on non-toxic chemical elements (Ti, Mo, Zr, Ta).
3. Elaboration of TMZT alloys
Taking into account the properties of the selected chemical elements, six variants of TMZT alloys were developed by vacuum arc remelting MRF ABJ 900 installation.
4. Preparation of alloys
The preparation of the alloys after solidification consisted in cutting them by electroerosion in order to obtain samples with standardized specific dimensions for each test and grinding them on abrasive paper, having different granulations in order to obtain a surface without scratches and cleaned by impurities.
5. Characterization of TMZT alloys
The study of the properties of TMZT alloys consists in laboratory investigations that determine the properties necessary for biomaterials through: chemical composition analysis, structural characterization, mechanical characterization, corrosion resistance and cytotoxicity assessment.
6. Discussion and interpretation of the results
The analysis of the results consisted in the graphic representation of the experimental values obtained and their interpretation compared to the results in the literature.
7. Conclusions
The final conclusions are the synthesis of the main results obtained, expressed in a concise and concrete manner, by highlighting the personal contributions and expressing some tendencies and perspectives of the subsequent researches on the finalization of the monography.

The program of experimental researches ensures the achievement of the main objective of the doctoral thesis and brings to the present the results with a significant contribution in the field of materials.

III.2. Methodology of Experimental Research

The theme improves the Ti-Mo alloys by: adding biocompatible elements, increasing the performance related to their mechanical properties and removing the disadvantages of the alloys currently used.

Experimental tests aim to characterizing TMZT alloys developed by chemical, structural, thermal, surface, mechanical and cytotoxic analysis. Types of investigations performed on the experimental materials are presented in Table III.2.

Table III.2. Characterization methods for elaborated experimental materials.

Alloys characterization	Method
Elementary composition	*Microanalysis with X-ray energy dispersion (EDX)*
Structural characterization	*Optical microscopy*
	X-ray Diffraction (XRD)
Thermal analysis	*Differential Scanning Calorimetry (DSC)*
Mechanical characterization	*Hardness tests*
	Indentation test
	Tensilestrengh test
	Compression test
	Fractographic analysis
Corrosion resistance	*Linear and cyclic polarization*
Surface characterization	*Contact angle (θ)*
Cytotoxicity assessment	*The MTT test (Tetrazolium Salt Method)*

The following laboratory investigations are considered:

- Elementary composition - it is necessary to determine the percentages of the chemical elements from elaborated TMZT alloys;

- Structural characterization - it is necessary to study microstructure, crystallographic orientation, texture and identification of constituent phases;

- Thermal analysis - provides information on the thermal behavior of alloys, especially in the temperature range in which they are used;

- Mechanical characterization - highlights the mechanical properties of elaborated TMZT alloys: hardness, modulus of elasticity, mechanical strength, elongation etc.;

- Corrosion resistance - determines the stability of alloys proposed in the human fluid;

- Surface characterization - involves measuring the contact surface of the alloy surface to achieve / optimize cell adhesion and proliferation;

- Cytotoxicity assessment - it is necessary to determine the processes taking place at the material interface - biological environment.

In this chapter are described analysis methods and equipment used in the experimental program.

The elaborated TMZT alloys were analyzed by structural, thermal, surface characterization, mechanical, corrosion resistance and cytotoxicity assessment methods. The products after casting were cut by electroerosion to obtain standard sample sizes required for each laboratory investigation.

In order to prepare the metallographic samples, a succession of stages was carried out as follows:

- The polishing process was done with polish paper. This is accomplished by successive grinding operations on large grain papers to small grain papers.

- The metallographic attack aims at highlighting the structural constituents. According to the ASM Handbook, the reactive used for titanium alloys has the following chemical composition: 10 ml of HF, 5 ml of HNO_3, 85 ml of H_2O, requiring a 30 s immersion time [13, 14].

The microstructure of alloys used in medical applications is defined by grain morphology, crystallographic orientation, texture, distribution, size and number of existing phases [15, 16].

III.2.1 X-ray Microanalysis with Energy Dispersion of X radiation

A complete characterization of a metallic material consists in knowing its composition, the concentration of the various elements or the impurities in the mass of the alloy. An extremely important aspect is the determination as precisely as possible, of the chemical composition of the TMZT alloys obtained after elaboration.

The EDAX system is a microanalysis detector, equipped with an electron microscope, which uses the resulting X-ray energy on the surface of the samples.

Determination of chemical composition can be performed, both punctually and in a well-defined region on the surface of the analyzed sample [17, 18].

This method is a variant of X-ray fluorescence spectroscopy, in which the sample investigatioon is based on the interactions between electromagnetic radiation and the sample (Figure III.1), analyzing the X radiation emitted by the sample as a response to the charging of particles loaded with electric charges. The characterization possibilities are largely due to the fundamental principle that each chemical element has a unique atomic structure that allows the X-rays characteristic of the atomic structure of an element to characterize it uniquely from another.

EDX

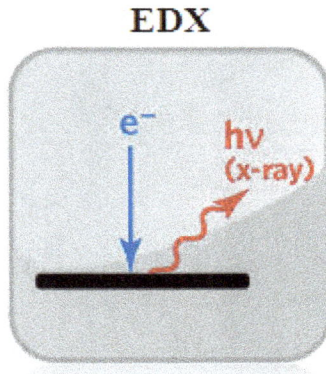

Figure III.1. Emission scheme of electrons on the surface of the analyzed sample [19].

To determine the chemical composition of alloys obtained from the TMZT system, the EDAX Bruker detector is attached to the electronic scanning microscope (SEM) Vega Tescan LMH II.

III.2.2 Analysis by Optical Microscopy

Optical microscopy provides detailed images of morphology and grain size and / or metallographic constituents with magnification up to 1000X.

Microstructural characterization by optical microscopy consists of:

• identification of structural alloys, their size, shape and distribution;

• microscopic determination of non-metallic inclusions in metals and alloys (sulphides, oxides, silicates, nitrides);

• the study of thermal and thermochemical treatment structures: new phases formed, defects of treatment, depth of the structure's modification;

• modification`s study of crystalline grains by plastic deformation;

• determination of crystallinity morphology and their distribution by size;

• the study of solid phase transformations according to temperature, etc. [20].

Optical microscope is a device that works on the principle of light reflection on a flat surface that allows clear images to be obtained through contrast and brightness settings. Structure investigation equipment is provided with several objectives, with different magnification powers. The operating principle of an optical microscope is shown in Figure III.2 [21].

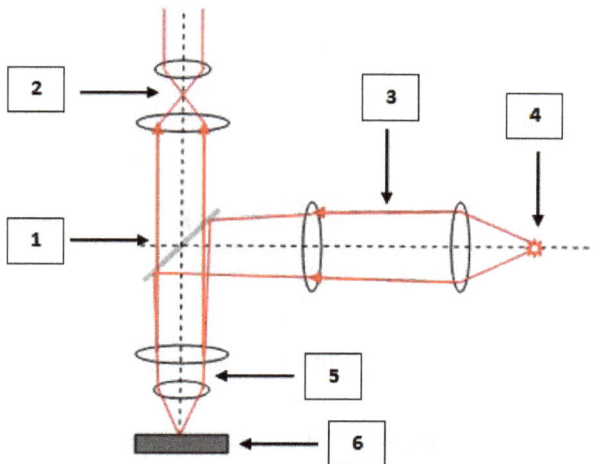

Figure III.2. Operation principle of an optical microscope: 1-axis reflector, 2-ocular; Condensation system; 4 light source; 5-object; 6 - sample [20].

For studying the structure of metals and alloys, different types of microscopes can be used to highlight constituent details. To evaluate the microstructures characteristics of the TMZT experimental alloys, an OPTIKA XDS-3 MET microscope equipped with an OPTIKAM 4083.B5 digital camera and the OPTIKAM B5.

III.2.3 Electronic Microscopy Analysis (SEM)

The scanning electron microscope is a device that produces high magnification images using electrons emitted by a tungsten filament to obtain the image.

The scanning electron microscope generally has the following components: electron gun, two or more magnetic lenses, tungsten filament encapsulated in a device called "Wehnelt", diaphragms, two scanning lenses, stigmatized lenses, one or more signal detectors, and electronics to process specimens collected from the sample and image formation. The schematic diagram of the electronic microscope is shown in Figure III.3 [21].

The main advantage of an electronic microscope is that it can produce images at high magnification powers of up to 50.000X on the surface of samples of different categories using HIGH VACUUM work mode.

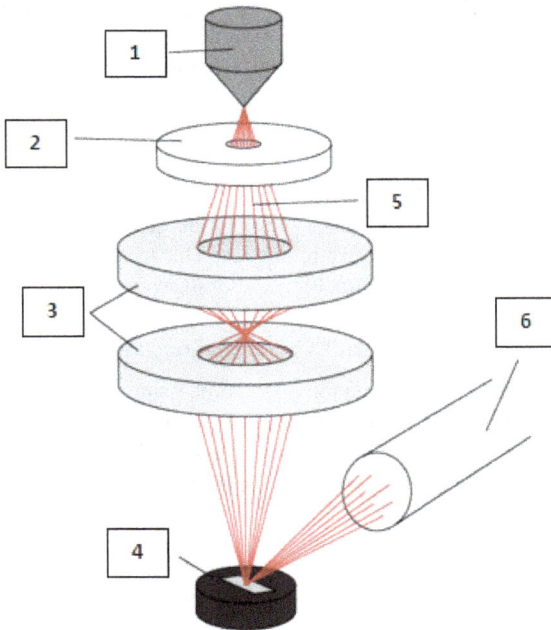

Figure III.3. Schematic diagram of an electronic scanning microscope: 1-gun electron, 2-anode, 3-lens magnetic, 4-sample, 5-beam electron, 6-detector [22-24].

Electron microscopy analysis to identify the microstructure and constituents was performed using a scanning electron microscope model Vega Tescan LMH II.

III.2.4 X-ray Diffractometry

X-ray diffractometry is a method of non-destructive analysis, being used in the study of materials for the determination of the crystalline structure by measurements on the symmetry and dimensions of the crystalline network, as well as by determining the occupied place of atoms in the elemental cell. X-ray diffractometry is also used to study the imperfections of the crystalline network, the magnitude and form of coherent spreading domains, and distortions within them [20, 25].

The most important applications of X-ray diffractometry aim to identify the phases and constituents of the material volume and the imperfections of the crystalline network. At the same time, it allows the study of crystallographic texture, internal stresses, measurements on interplanar distances (Bragg method), thin-film diffraction on thin layers, etc. [17, 26, 27]. For structural X-ray analysis, a PANalytical X'Pert PRO MPD X-ray diffractometer was used.

III.2.5 Differential Calorimetric Analysis

Differential Calorimetric Analysis (DSC) refers to the quantitative measurement of heat energy exchange during a thermodynamic process. This method is used to determine the following: transition temperature; melting / boiling point; crystallization time and temperature; degree of crystallinity; dissolution heat and change during reactions; specific heat; oxidative / thermal stability; kinetics of reactions; purity of samples [6, 26].

The DSC allows obtaining of information on the processing temperatures associated with the alloying processes. The results of the calorimetric measurements are represented by a measurement curve termed a thermogram, conventionally denominated in the DSC curve. The general scheme of a DSC is shown in Figure III.4 [6, 89].

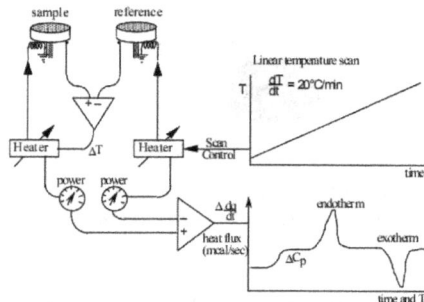

Figure III.4. The general scheme of a DSC [28].

Both the sample crucible and the standard (which is an empty crucible) are placed inside the DSC system cell. The temperature difference between the sample and the reference is measured and recorded as a heat flux.

The temperature is measured using three thermocouples, one indicating the sample temperature, the second reference temperature, and the last furnace temperature. The output signal from DSC is taken over by a computer that performs data analysis by means of suitable software (NETZSCH PROTEUS Thermal Analysis). DSC analysis equipment also includes a cylinder with inert gas (argon) and one with coolant (liquid nitrogen) [6].

The characteristics of the calorimeter used are as follows:

• Temperature range: -170 ... + 600°C;

• Heating speed: 0.001 K / min ... 100 K / min;

• Cooling speed: 0.001K / min ... 100K / min;

• Temperature precision: 0.1 K;

• Sensitivity: <1 µW;

• Precision of enthalpy: ± 0.5%;

• Argon atmosphere [29].

For Differential Calorimetry, the Differential Scanning Calorimeter (DSC) type F3 Maia (NETZSCH) was used in the argon protective atmosphere.

III.2.6 Determination of Hardness

Hardness is a property of materials that express their ability to resist the action of mechanically penetrating a tougher body into its surface. When determining the hardness of the materials, the size of the traces produced by a penetration body, characterized by a certain shape and size and the force acting on it, are taken into account [30].

The methods for determining the hardness, depending on the speed of the force on the penetrator, are classified in static methods, where the drive speed is below 1 mm / s, and dynamic methods for which the drive speed exceeds this value [31].

The Vickers Hardness Determination Method (Figure III.5) uses as a penetrator a diamond in the form of a pyramid with a square base, and consists in pressing it at a reduced speed and with a certain predetermined force F on the surface of the test material. The Vickers hardness, symbolized by HV, is expressed by the ratio of the applied force F to the area of the lateral surface of the residual trace produced by the penetrator. The trace is considered to be a straight pyramid with a square base, with diagonal d, having the same angle as the penetrator at the top [26].

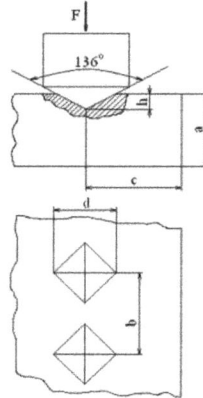

Figure III.5. Vickers hardness scheme [27].

For the Vickers hardness determination method, at least three attempts are made on the test material. For each trace the average diagonal value is calculated based on the magnitude of the two diagonals measured. It is recognized that the difference in diagonal dimensions is within an error margin of not more than 2%.

The hardness measurements highlight resistance and provide pieces of information on the behavior of the studied materials. In this way, we can analyze TMZT alloys developed for the purpose of fitting them into a specific medical application [32, 33].

HV hardness measurements on TMZT alloys were performed on Wilson Wolpert 751 N.

III.2.7 Microindentation Method

The measurement of the longitudinal elastic modulus for the obtained TMZT alloys was achieved by the microindentation method. This method consists of penetrating the surface of the sample with a conical palpate at a certain force.

During the microindentation test, the values of the loading forces are recorded relative to the penetration depth of the indenter in the material layer. Figure III.6 shows the principle of measuring the microindentation test.

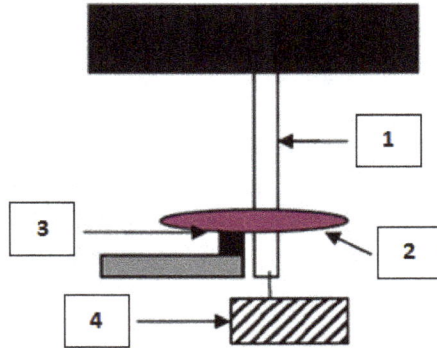

Figure III.6. Microindentation test measurement principle: 1 - indenter, 2 - disc reference, 3 - capacitive sensor, 4 - sample [34].

To perform the microindentation test on the tribometer, place the linear mass on which the sample is fixed, attaching to it the capacitive sensor of the penetration depth sensor in the test material. A force sensor is used on which the rigid support in which the indentor support system is mounted by means of two screws. Between the spindle support system and a metal disc, called the reference disc, under which the capacitive sensor is placed to record the vertical displacement of the indentation (indentation depth) [35-36] is fixed.

To perform this indentation test, it was necessary to use the CETR UMT-2 microtribrometer.

III.2.8 Tensile Strength Test

The tensile strength test represents the basic test of a material for the purpose of determining its main mechanical properties. The test involves stretching samples of different sections by progressive and continuous application of a load and without shocks in the direction of the longitudinal axis, generally to breakage, while measuring the deformations corresponding to the various tensile forces values. During the test, the sample is stretched, snapped, harden, and finally breaks [10].

Samples subjected to tensile strength are of standardized form and size and must meet certain conditions:

• the dimensions of the sample are large enough so that the results are not influenced by the particularities of the behavior of some crystalline formations of the material and that the elongation can be measured with sufficient precision;

• to exist in a particular area of the sample, a homogeneous tension , the local tensions appearing in the specimen gripping portions, must to be minimal and not to affect the stress state of the main sample area [30, 31].

The INSTRON 8801 Servo Mechanical Testing System provides complete static and dynamic test solutions for both metallic and non-metallic materials of standard materials and advanced materials with testing on. The INSTRON 8801 has a loading capacity of up to 100 kN, a working space between large sleepers, high stiffness and good alignment accuracy.

Moreover, with precision mechanical systems that are combined with advanced Dynacell 8800 ™ digital controller and Dynacell power cell functions, the Instron 8801 meets the most demanding dynamic and static test demands. The universal static and fatigue test machine INSTRON 8801 is distinguished by the stability of induced size, force, displacement and deformation [37].

The traction test was carried out in compliance with the national and international standards in force, using standardized size samples by applying a progressive tensile strength load in the longitudinal axis direction. Thus, a number of mechanical characteristics have been determined which allow the assessment of the behavior of the alloys undergoing, during their application [10, 38-40].

For each test, rectangular cross-sectional samples of flat shape and specific dimensions were made. Samples were obtained by cutting on a wire electroerosion machine [10].

The tensile test determinations on samples from the TMZT alloy system were performed on the INSTRON 8801 Servo System Testing System.

III.2.9 Compression Test

The compression test is a destructive testing of materials for the purpose of experimentally determining the main strength and plasticity characteristics of the materials. For this purpose, compression tests have been carried out on samples from elaborated TMZT alloys, cut to specific dimensions.

The compression test is performed by placing the test piece between the platters of a universal machine to be tested. In the compression test two phenomena characteristic of this test occur: the buckling phenomenon (loss of stability) occurring at the compression of the long bars: l> 5d, where l is the length of the bar, and d is the cross sectional dimension and the phenomenon of "barrels" due the friction of ends of the sample and the test machine platters (Figure III.7) [40, 41].

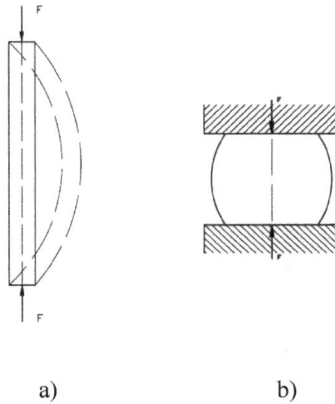

a)   b)

Figure III.7. Characteristic phenomena appearing for the compression test: a) buckling, b) barrels [39].

Another problem specific to the compression test is that, in the case of tenacious materials, cannot be broken: they deform plastic continuously as the load increases.

The compression test machine used to test the TMZT alloys was Walter + Bai AG.

III.2.10 Fractographic Method

Fractographic analysis consists in analyzing the tear-off surface, made with the naked eye, with the usual microscope or electron microscope. For the proper design and use of metals, it is important to know their behavior before breakage, and the conditions in which it occurs [10].

By analyzing the shape and appearance of the samples, we differentiate different ways of breaking:

• Ductile rupture, preceded by appreciable deformations;

• Fragile rupture, preceded by very small deformations;

• Mixed rupture (ductile and fragile).

In case of ductile rupture, the material suffers strong plastic deformation, highlighted by considerable values of elongation and strangulation at break (Figure III.8 c, III.8 a). The materials with ductility-in breakage have shown an important bottleneck and also a proportional elongation. In this case, the rupture propagates from the center of the section to the directions of the maximum tangential stresses, at 45° angles, the appearance of the breaking surface being characteristic, called the cone (Figure III.8b, III.8a).

Fragile tearing produces cuts in a plane approximately perpendicular to the stress plane and has a crystalline structure (rupture starts at grain boundaries). Factors that cause different tensile strength results and produce apparent changes in the mechanical strength characteristics are: sample sizes, load speed and test machine characteristics.

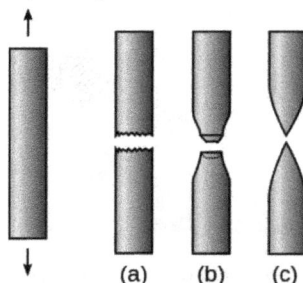

Figure III.8. Breaking modes of the materials: a) fragile, b) mixed, c) ductile [42].

From a danger point of view which it presents, ductile behavior is preferred, allowing a redistribution of stresses that cause rupture, a cancellation of stress peaks (in concentrators), avoiding catastrophic rupture [39, 43 - 46].

The appearance of the breakage surfaces of TMZT alloys was analyzed by metallographic study using electron microscopy.

Samples of experimental TMZT alloys subjected to break at tensile strength test were studied by fractographic analysis.

For the analysis of the breaking surfaces, an electronic microscope Quanta Inspect S, FEI (The Netherlands) was used.

III.2.11 Characterization of Electrochemical Behavior

Corrosion represents the phenomenon of partial or total destruction of materials and especially of metals due to chemical or electrochemical reactions to interaction with the environment or with specific environments, being one of the most important parameters that determine the functionality of a biomaterial [47, 48].

When implanting a biomaterial into the body, corrosion can lead to the formation of reaction products or metal ions that can damage the health of the body. Releasing elements from biomaterial can cause allergic reactions and damage to adjacent soft tissues. Most metals used in medical applications such as Fe, Cr, Co, Al, Ni, Ti, Ta, Mo and W are tolerated by the body if they are in well-determined amounts, being even essential in metabolic processes [49, 50].

The physiological environment is very aggressive and therefore the corrosion resistance of the implanted metal materials is considered to be an important aspect of their biocompatibility. The extracellular fluid, which is most often in contact with the implants, contains large amounts of ions such as: Na +, Cl-, HCO3-, OH- and reduced amounts of other ions, with a pH around 7.4, maintained at a constant temperature of 37°C [51].

To calculate the corrosion rate of an immersed alloy in a corrosive environment, it is necessary to know the instantaneous current density determined by the polarization resistance method. This method is used to determine the corrosion current to the corrosion potential of a metal or alloy using for this purpose the linear polarization curve obtained for relatively small overvoltage. The corrosion current determined by this path represents the current occurring at the metal / medium interface when the metal is immersed in solution and can be measured directly by electrochemical methods. The schematic diagram of corrosion investigation equipment is shown in Figure III.9.

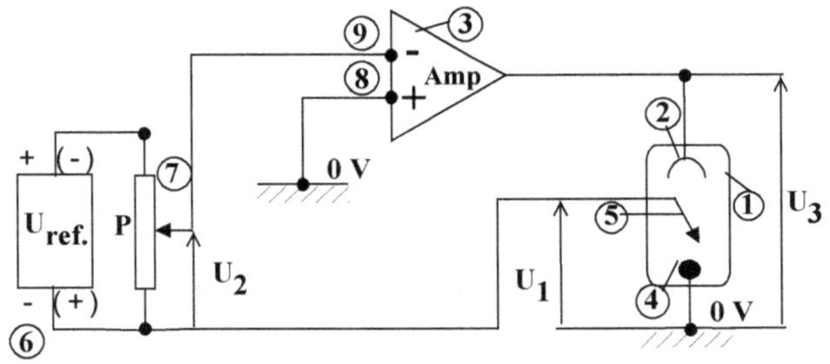

Figure III.9. Schematic diagram of an electrochemical corrosion investigation equipment: 1- electrochemical cell, 2 auxiliary electrode; 3 - amplifier, 4 - working electrode, 5 reference electrode, 6 - reference source, 7 - multiturn potentiometer, 8,9 - power [52].

Electrochemical corrosion study at laboratory level is done through specific tests such as linear and cyclic polarization, which will determine the behavior of the alloys in the corrosion environment.

Cyclic voltammetry is an electrochemical method by which we can study the mechanisms of electrochemical reactions. The resulting experimental curves are called voltamograms. They use a DC signal and are cycled. In a cyclic voltammetry experiment, the potential is scanned between a positive maximum and another negative backward velocity, with a constant scanning rate. Through the tracing of the Tafel curves, parameters that characterize the corrosion resistance of the TMZT samples can be determined: the corrosion potential (Ecorr), the slope of the cathode curve (βc), the slope of the anode curve (βa), the density of the corrosion current (jcorr) Jpass), corrosion rate (Vcorr).

In order to determine the corrosion potential and plot voltagrams, a Voltech Lab 21 Economical Potentiostat electrochemical system it used.

III.2.12 Contact Angle Measurement (θ)

One of the requirements of biomaterials is cellular adhesion on the surface of the material, depending on surface energy. Specialty studies in domain indicated that contact angle measurement is important for the study of cell adhesion to the surface of the studied material [6, 53].

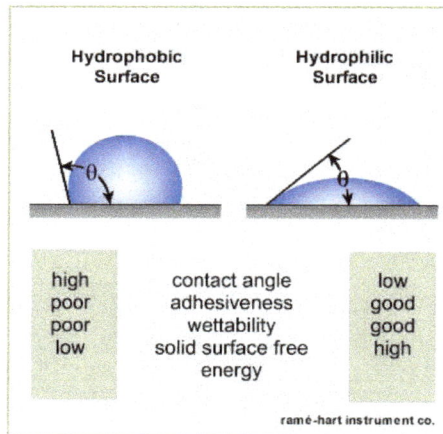

Figure III.10. Measurement of contact angle for hydrophilic and hydrophobic surfaces [55].

Measurement of the contact angle (Figure III.10) is an experimental technique used to evaluate the hydrophilic or hydrophobic character of the surfaces. Surfaces can be classified as hydrophilic or hydrophobic reported at 90°. If the angle of contact is between 0-90°, the material is hydrophilic and if the angle of contact is between 90-180°, material is hydrophobic [6, 54].

The choice to use distilled water as a blank for assessing contact angle for experimental TMZT alloys, is based on the fact that water represents 70% of the human body.

The equipment used allows the determination of the surface tension of the liquids as well as of the free surface energy of the solid. The principle of measuring the angle of contact (Figure III.11) consists in placing a drop of water with a microsurge with the drop volume of 4 microlitres. Drop lighting is made from behind and recorded from the opposite side with a digital camera. The image obtained is further analyzed through the Famas program, a KYOWA integrated goniometer software [6].

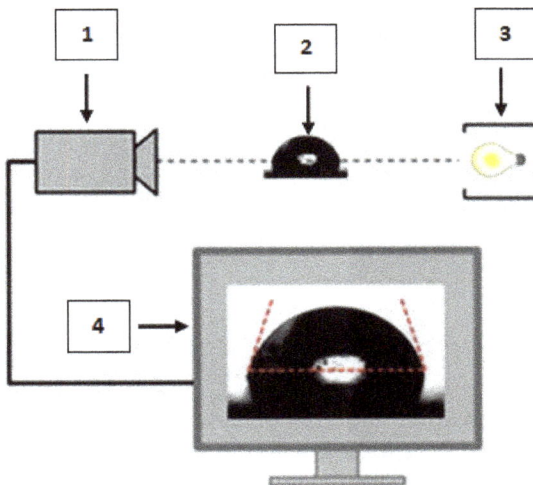

Figure III.11. The principle of measuring the contact angle: 1-video camera, 2-sample, 3-light, 4-computer and screen [56].

For measure the contact angle, the Kyowa DM-CE 1 was used.

Titanium-Based Alloys for Biomedical Applications   Materials Research Forum LLC
Materials Research Foundations **74** (2020)   https://doi.org/10.21741/9781644900796

III.2.13 Evaluation of Cytotoxicity

Generally, metals in contact with the biological body give complex effects, producing a series of biological reactions, depending on the concentration, the exposure time, etc., the reciprocal reaction of the biomaterial-organism being beneficial or harmful.

Biocompatibility involves accepting a material by surrounding tissues, implicitly by the whole organism [57].

The use of cell cultures in biocompatibility testing has a number of advantages:

• high sensitivity of culture cells to toxic agents;

• the possibility of investigating specific interactions at the cellular or molecular level;

• Possibility to carry out extensive series of experiences under the same conditions [58].

Testing the cytotoxicity of materials is one of the important and necessary tests to which materials are subjected to determine the compatibility of materials with human tissue.

For TMZT alloys developed, biocompatibility will be highlighted by the MTT (Tetrazolium Salt Method) test. This technique has been chosen because studies [59-63] have shown that the MTT test is mostly used in biomaterial evaluation.

Cell viability was assessed by the MTT test, which allows quantification of the live cell number. The MTT test on the TMZT alloys was conducted through a series of steps, the experiment being performed according to appropriate procedure standards and procedures.

The samples belonging to each TMZT alloy were prepared by cutting to set dimensions, followed by their exposure for 30 minutes on each side to the UV action in a transiluminator.

Samples were plated into wells plates for seeding with cell cultures. Removal of the culture medium from the wells was performed by aspiration and the cells were washed with sterile PBS (Phosphate Buffered Saline Solution).

125µL of MTT solution was added to the wells, followed by incubation of the plate at 37° C (95% humidity, 5% CO2) for 3 hours. After 3 hours, the supernatant from the wells was aspirated and 100 µL DMSO (dimethyl sulfoxide) per well was added to make cell lysis to solubilise the intracellular formazan crystals.

After shaking the plate for 5 minutes on an orbital shaker, plate assay was performed using a spectrophotometer (Triad LT). In order to obtain a quantification of the amount of formazan that is directly proportional to the number of viable cells, the evaluation was performed at two incident wavelengths, $\lambda = 630$ nm (test) and $\lambda = 570$ nm.

Finally, cell viability was expressed as a percentage by reference to the control wells (blanks with no cells, Control - wells with culture medium and cells) according to the formula:

$$\% \ CV = 100 \times (ODs\text{-}ODb)/(ODc\text{-}ODb) \hspace{3cm} (3.1)$$

In which:

• ODs is the optical density (in units) for the sample;

• ODb - optical density for cell-free wells (blank);

• ODc - optical density for control wells [112, 113].

Experimental investigation by MTT test of TMZT alloys, provided the following equipment: laminar flow hot - Safeflow 1.2; Spectrophotometer (Triad LT); centrifuge with cooling - Jouan MR 18-22; CO_2 incubator, water jacket and double enclosure - ThermoScientific and Olympus IX51 phase contrast fluorescence and contrast microscope.

The elaborated TiMoZrTa alloys (TMZT) are intended to be used in medical applications to meet functional requirements and to remove the disadvantages of classical alloys used as biomaterials in human tissue.

The use of metallic biomaterials in the human body is dependent on certain specific characteristics of the material as well as on the specific function it has to perform.

The superior utilization of metallic biomaterials suppose the knowledge of chemical, physical, mechanical, thermal, electrical, magnetic, optical, and technological properties by means of specific methods of these materials.

The experimental research program described aims at analyzing the TMZT alloys developed by laboratory investigations using performant equipment, for chemical, structural, physico-mechanical and cytotoxic to a complete characterization of the samples and defining these alloys as materials compatible with human tissue. The equipment used to complete the experimental research program is presented in Table III.3

Table III.3. Equipment used.

Investigation name	Samples used	Equipment used for alloys	Equipment name
Equipment used for the alloys obtaining			
Elaboration of alloys in the TMZT system			Vacuum Arc Retrieval System MRF ABJ 900
Equipment used for the alloys preparing			
Alloys cutting			Japax Japt machine
Embedding samples			Metapress-M machine
Metallographic preparation of samples - grinding, polishing			Grinding and polishing machine - Force 2V
Equipment used for chemical, structural and thermal characterization of TMZT alloys			
X-ray microanalysis with X-ray energy dispersion (EDX)			Scanning Electron Microscope, model Vega Tescan LMH II
Optical microscopy for structural analysis			OPTIKA XDS-3 MET Microscope

Investigation name	Samples used	Equipment used for alloys	Equipment name
X-ray Diffraction for Qualitative Phase Analysis (XRD)			X-ray Diffractometer - X'Pert Pro MPD
Thermal analysis			Differential scanning calorimeter type F3 Maia
Equipment used for physico-mechanical characterization of TMZT alloys			
Determination of hardness			Wilson Wolpert 751 N
Indenting tests			Microtribrometer Cetr-UMT-2

Investigation name	Samples used	Equipment used for alloys	Equipment name
Tensile strength test			Hydraulic test system 8801
Compression test			Walter+Bai AG testing machine
Equipment used for corrosion resistance analysis			
Determination of corrosion rate by electrochemical methods			Volta Lab 21potentiostat

Equipment used for surface characterization			
Contact angle			Kyowa DM-CE 1contact angle measuring equipment
Equipment used for cytotoxicity assessment			
MTT test			Safeflow 1.2
			Spectrophotometer (Triad LT)
			Cooling Centrifuge - Jouan MR 18-22
			CO$_2$ incubator, water jacket model and double enclosure – ThermoScientific

The analyses and laboratory methods used in the book have enabled the identification of the necessary characteristics (structural, mechanical, electrochemical, cellular viability) and the evaluation of the TMZT alloys developed for their use as materials with applicability in the medical field.

References

[1]   W.F. Ho., C.P. Ju, J.H. Chern Lin, Structure and properties of cast binary Ti-Mo alloys, Biomaterials, 20 (1999) 2115-2122. https://doi.org/10.1016/S0142-9612(99)00114-3

[2]   X.H. Min, S. Emura, L. Zhang, K. Tsuzaki, Effect of Fe and Zr additions on ω phase formation in β-type Ti–Mo alloys, Materials Science and Engineering A, 497 (2008) 74–78. https://doi.org/10.1016/j.msea.2008.06.018

[3]   Y. Li., C. Wong, J. Xiong, P. Hodgson, C. Wen C., Cytotoxicity of titanium and titanium alloying elements, Journal of Dental Research, vol. 89 (5), 2010, pp. 493-497. https://doi.org/10.1177/0022034510363675

[4]   M. Gómez-Florit, J.M. Ramis, R. Xing, S. Taxt-Lamolle, H.J. Haugen, S.P. Lyngstadaas, et al., Differential response of human gingival fibroblasts to titanium- and titanium-zirconium-modified surfaces, J. Periodontal Res., vol. 49(4), 2014, pp. 425–436. https://doi.org/10.1111/jre.12121

[5]   M.S. Cercel (cas. Bălțatu), Contribuții privind îmbunătățirea proprietăților aliajelor de Ti-Mo destinate aplicațiilor medicale – teză de doctorat, Iași, 2017.

[6]   M.S. Baltatu, P. Vizureanu, T. Balan, M. Lohan, C.A. Tugui, Preliminary Tests for Ti-Mo-Zr-Ta Alloys as Potential Biomaterials, Book Series: IOP Conference Series-Materials Science and Engineering, 374 (2018), 012023. https://doi.org/10.1088/1757-899X/374/1/012023

[7]   G.T. Pop, Biomateriale si componenete protetice metalice, Ed. Tehnopress, Iași, 2004.

[8]   D.M. Bombac, M. Brojan, P. Fajfar, F. Kosel, R. Turk, Review of materials in medical applications, Materials and Geoenvironment, vol. 54(4), 2007, pp. 471-499.

[9]   M. Geetha, A.K. Singh, R. Asokamani, A.K. Gogia, Ti based biomaterials, the ultimate choice for orthopaedic implants - A review, Mater. Sci., 54 (2009) 397-425. https://doi.org/10.1016/j.pmatsci.2008.06.004

[10] M.S. Bălțatu, P. Vizureanu, V. Goanță, C.A. Tugui, I. Voiculescu, Mechanical tests for Ti-based alloys as new medical materials, IOP Conf. Ser.: Mater. Sci. Eng., 572 (2019) 012029. https://doi.org/10.1088/1757-899X/572/1/012029

[11] J.C. Fanning, Titanium 95' science and technology; TIMETAL21SRx., 1996, pp.1800-1807.

[12] A.K. Mishra, J.A. Davidson, P. Kovacs, R.A. Poggie, Beta titanium in the 1990s. Warrendale, Pennsylvania: The Mineral, Metals and Materials Society, 1993, pp. 61–66.

[13] ***ASM Handbook, Alloy Phase Diagrams, vol.3.

[14] *** ASM Handbook, Metallography and Microstructure, vol.9.

[15] F. Miculescu, Tehnici de analiză și control a biomaterialelor, Ed. Printech, București, 2009.

[16] J.R.S.M. Júnior, R.A. Nogueira, R.O. Araújo, T.A.G. Donato, V.E.A. Chavez, A.P.R. Claro, J.C.S. Moraes, et all, Preparation and Characterization of Ti-15Mo Alloy used as Biomaterial Materials Research, 14(1), 2011, pp. 107-112. https://doi.org/10.1590/S1516-14392011005000013

[17] I. Hopulele, N. Cimpoeșu, C. Nejneru, Metode de analiză a materialelor. Microscopie și analiză termică, Ed. Tehnopress, Iași, 2009.

[18] C. Munteanu, M. Ștefan, C. Baciu, N. Cimpoeșu, Metode difractometrice și microscopie optică și electronică în studiul materialelor, Ed. Tehnopress, Iași, 2008.

[19] Information on www.eag.com

[20] P. Vizureanu, Metode și tehnici de cercetare în domeniu – Studiul materialelor prin microscopie optică, Universitatea Tehnică "Gheorghe Asachi" din Iași, 2011.

[21] D.G. Galușcă, C. Nejneru, M.C. Perju, D.C. Achiței., Tehnologii de tratare a suprafețelor metalice, Straturi subțiri obținute prin depunere, Ed. Tehnopress, Iași, 2012.

[22] Information on www.eng-atoms.msm.cam.ac.uk/RoyalSocDemos/SEM

[23] Information on www.fei.com

[24] *** Fei Company, Quanta 200 3D-User's Operation Manual - Third Edition, 2004

[25] M. Bibu, Metode și tehnici de analiză structurală a materialelor metalice, Ed. Universității „Lucian Blaga", Sibiu, 2000.

[26] I. Rusu, Tehnici de analiza in ingineria materialelor, Ed. PIM, Iași, 2011.

[27] V. Popovici, A.C. Pavalache, I.M. Vasile, I. Voiculescu, E.M. Stanciu, D. Pausan, Finite Element Method for Simulating the Vickers Hardness Test, Applied Mechanics and Materials, 555 (2014) 419-424. https://doi.org/10.4028/www.scientific.net/AMM.555.419

[28] Information on http://www.colby.edu/

[29] Information on ***https://termoanalitic.wordpress.com

[30] A. Atanasiu, T. Canta., A. Caracostea, I. Crudu, I. Drăgan, et all, Încercare materialelor, vol. 1, Ed. Tehnică, București, 1982.

[31] Zecheru Gh., Drăghici Gh., Elemente de științe și ingineria materialelor, vol. 1 și 2, Ed. ILEX și Ed. Universității din Ploiești, 2001.

[32] Information on *** www.wolpertgroup.com

[33] ***Manual de utilizare,Wilson Wolpert 751 N, WOLPERT Group, Universal Brinel, Wilson Instruments, An Instron Company.

[34] Information on *** http://www.cetr.com

[35]   Information on *** http://www.mec.tuiasi.ro

[36]   C.A. Schuh, Nanoindentation studied of materials. Material Today, vol. 9(5), 2006, pp. 32-40. https://doi.org/10.1016/S1369-7021(06)71495-X

[37]   Information on www.instron.us

[38]   ***SR EN ISO 6892-1:2010.

[39]   G. Buzdugan, Rezistenţa materialelor, Ed. Academiei Republicii Socialiste Romania, Bucureşti, 1986.

[40]   V. Goanţă, Mecanica ruperii, Ed.Tehnopress, Iaşi, 2006.

[41]   I. Voiculescu, O. Donţu, V. Geantă, Ganatsios S., Effect of the Laser Beam Superficial Heat Treatment on the Gas Tungsten Arc Ti-6Al-V Welded Metal Microstructure. Proceedings of the Society of Photo-Optical Instrumentation Engineers (SPIE), Conference on Industrial Laser Application (INDLAS), Art. No. 70070M, IDS Number: BIF91, Bran, 2007. https://doi.org/10.1117/12.801972

[42]   N. Geru, Materiale metalice. Structura, Proprietati, utilizări, Ed.Tehnică, Bucureşti, 1985.

[43]   I. Voiculescu, C. Rontescu, I.L. Dondea, Metalografia îmbinărilor sudate, Editura Sudura, Timişoara, 2010.

[44]   I. Voiculescu, I.M. Vasile, E.M. Stanciu, A. Pascu, Ştiinţa şi ingineria materialelor, Indrumar de laborator, Ed. Lux Libris, Braşov, 2015.

[45]   ***Carte tehnica: Microscop electronic Quanta Inspect S, Laborator LAMET, CT – ME – UPB-SML-05.01/rev0/07, Bucuresti, 2007.

[46]   L. Breteanu, Rezistenţa materialelor- îndrumar de laborator, Ed. Univ., Tg. Mures, 1995.

[47]   N.T.C. Oliveira, A.C. Guastaldi, Electrochemical behavior of Ti-Mo alloys applied as biomaterial, Corrosion Science, 50 (2008) 938-945. https://doi.org/10.1016/j.corsci.2007.09.009

[48]   M.I. Popa, D. Mareci, Electrochimie generală şi coroziune, Editura Politechnium, Iaşi, 2009.

[49]   M. Gómez-Florit, J.M. Ramis, R. Xing, S. Taxt-Lamolle, H.J. Haugen, S.P. Lyngstadaas, et al., Differential response of human gingival fibroblasts to titanium- and titanium-zirconium-modified surfaces, J. Periodontal Res., vol. 49 (4), 2014, pp. 425–436. https://doi.org/10.1111/jre.12121

[50]   Rusu L., Implante chirurgicale. Studii şi cercetări în vederea omologării-teză de doctorat, Universitatea "Politehnica", Timişoara, 2006.

[51] N.T.C. Oliveira, G. Aleixo, R. Caram, A. Guastaldi, Development of Ti-Mo alloys for biomedical applications: Microstructure and electrochemical characterization, Materials Science and Engineering A, 452 (2007)727-731. https://doi.org/10.1016/j.msea.2006.11.061

[52] Information on ***http://www.scritub.com

[53] E.A. Vogler, Structure and reactivity of water at biomaterial surface. Adv. Colloid Interface Sci., 74 (1998) 69-117. https://doi.org/10.1016/S0001-8686(97)00040-7

[54] R.E. Baier, Surface behavior of biomaterials: the theta surface for biocompatibility, Journal of material science: Materials in medicine, vol. 17(11), 2006, pp. 1057-1062. https://doi.org/10.1007/s10856-006-0444-8

[55] ***http://www.ramehart.com/contactangle.htm

[56] ***https://aparatemasura.wordpress.com

[57] D.P. Balaban, Biomateriale, Ed. Ovidius University Press, Constanţa, 2005.

[58] F. Eiras, M.H. Amaral, R. Silva, E. Martins, J.M. Sousa Lobo, A.C. Silva, Characterization and biocompatibility evaluation of cutaneous formulations containing lipid nanoparticles, International Journal of Pharmaceutics, 519 (2017) 373–380. https://doi.org/10.1016/j.ijpharm.2017.01.045

[59] T. Mosmann, Rapid colorimetric assay for cellular growth and survival: application to proliferation and cytotoxity assays, J Immunol Methods, vol. 16(65), 1983, pp. 55-63. https://doi.org/10.1016/0022-1759(83)90303-4

[60] G. Ciapetti, E. Cenni, L. Pratelli, A. Pizzoferrato, In vitro evaluation of cell/biomaterial interaction by MTI' assay, Biomaterials, 14(5), 1993, pp. 359-36. https://doi.org/10.1016/0142-9612(93)90055-7

[61] D.F. Williams D.F., Biofunctionality and Biocompatibility, Materials Science and Technology, Vol. 14, Medical and Dental Materials, Weinheim, 1992, pp. 2-27.

[62] S.L. Cotescu, G. Jicmon, Aspecte ecologice ale utilizării titanului şi aliajelor sale pe scară largă, Simpozionul naţional"Impactul acquis-ului comunitar de mediu asupra tehnologiilor şi echipamentelor, Constanţa, 2008.

[63] A.I. Stavrakis, J.A. Niska, A.H. Loftin, L.M. Kwong, F. Billi, L.S. Miller, N.M. Bernthal, An in vivo Assessment of the Bacterial Susceptibility of Porous Tantalum, Orthopaedic Hospital Research Center, California, USA.

CHAPTER IV: Obtaining of New Ti-Based Alloys for Medical Applications

In the last years, the second generation biomaterials were appeared, which are synthesized considering the existence and control of the physical, chemical and biological processes at the implant / tissue interface, so that the normal cellular processes are stimulated.

The titanium has attracted the attention of the medical world through its particularly advantageous properties: biocompatibility, low thermal conductivity, low density, corrosion resistance, the cost price of the material being four times lower than that of gold [1, 2].

The composition, structure, state and mechanical characteristics are in the basis for obtaining titanium alloys.

IV.1 Design of TMZT Alloys

Designing and obtaining metallic materials commonly used as biomaterials requires a thorough study of both the design and the technological parameters of the developmental equipment.

At the basis of TMZT alloy design, stay information on the physical properties of alloying elements (melting point and density - Table IV.1), as well as the equilibrium diagrams for Ti-Mo alloys (Figure IV.1), Mo-Ti- Zr (Figure IV.2), Mo-Ta-Ti (Figure IV.3), Ta-Ti-Zr (Figure IV.4) [3-5].

Table IV.1. The physical properties of the alloying elements used for obtaining TMZT alloys [1, 6-7].

Chemical element	Melting point [°C]	Density [kg/m^3]
Titanium	1668	4507
Molybdenum	2625	10200
Zirconium	1855	6511
Tantalum	3020	16650

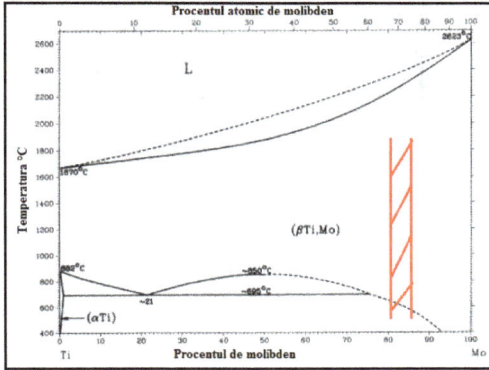

Figure IV.1. Binary Ti-Mo alloy phase diagram [3].

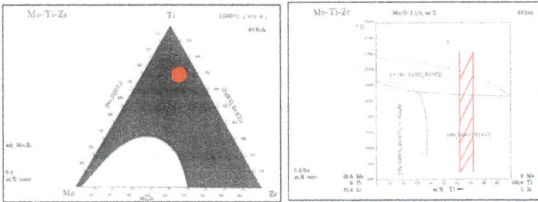

Figure IV.2. Phase diagrams for Mo-Ti-Zr alloy: a) ternary, b) of equilibrium [8].

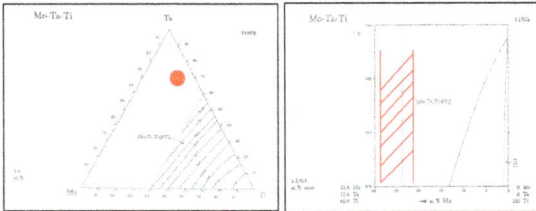

Figure IV.3. Phase diagrams for Mo-Ta-Ti alloy: a) ternary, b) of equilibrium [8].

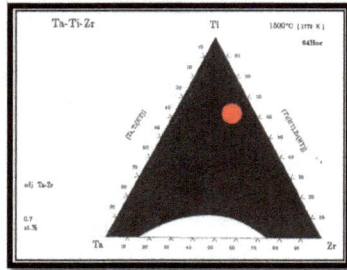

Figure IV.4. Ternary diagram for Ta-Ti-Zr alloy [8].

Figure IV.1 shows the binary thermodynamic equilibrium diagram of the Ti-Mo alloy system, where all the phase transformations that occur in the alloy system can be seen. This highlights the total liquid solubility of molybdenum in titanium.

The studied alloys are in the range of quaternary alloys (TiMoZrTa) with variations of alloying elements in the following range: 5-15% Ta, 15-20% Mo, Zr kept constant, the balance being achieved with Ti, which is the main element.

The field of alloying in the ternary diagrams (Figures IV.2, IV.3, IV.4) was represented in two ways: a colored area through a red disc in the case of isothermal sections or a hatched area on a vertical section (the edges they are the maxima and the minimum concentrations).

In the Mo-Ti-Zr ternary system liquid reactions occur (with the formation of a Mo2Zr intermetallic compound) of the type:

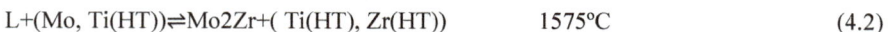

$$l \rightleftharpoons (Mo, Ti(HT), Zr(HT)) \qquad\qquad 1620°C \qquad\qquad (4.1)$$

$$L + (Mo, Ti(HT)) \rightleftharpoons Mo2Zr + (Ti(HT), Zr(HT)) \qquad 1575°C \qquad\qquad (4.2)$$

The diagrams from Figures IV.1, IV.2, IV.3, IV.4 represent the foundation for designing the proposed TMZT alloy system from the point of view of all phase transformations that occur at different temperatures and concentrations of the alloying elements.

Titanium alloy production processes require special working conditions such as:

• Melting and casting are not possible with the usual facilities of the laboratories, therefore they cannot carry out the heating process at 1800-2000°C and also cannot not provide a sufficiently efficient protective environment;

• Chemically, titanium alloys and pure titanium reacts at high temperatures with the gaseous elements from environment: O_2, H_2, N, therefore casting of these alloys must only be carried out in a controlled atmosphere (argon);

• Crucibles or casting molds must have the necessary refractoriness to prevent chemical reaction with titanium;

• Since high melt temperatures and metal densities can create problems with the chemical and structural homogeneity of the melted and solidified alloy, repeated remeltings have to be carried out for refining and homogenization [9, 10].

After consultation with the literature [9-31], six TMZT alloys with different percentages of the alloying elements were proposed and developed: Ti - 58 ÷ 73%, Mo - 15 ÷ 20%, Zr - 7% and Ta - 5 ÷ 15%. The alloys proposed for development are presented in Table IV.2 together with the alloying reports

Table IV.2. Alloys proposed for development.

Composition alloy				Alloying reports
Ti [%]	Mo [%]	Zr [%]	Ta [%]	
73	15	7	5	10.4 : 2.1 : 1 : 0.7
68	15	7	10	9.7 : 2.1 : 1 : 1.4
63	15	7	15	9 : 2.1 : 1 : 2.1
68	20	7	5	9.7 : 2.9 : 1 : 0.7
63	20	7	10	9 : 2.9 : 1 : 1.4
58	20	7	15	8.3 : 2.9 : 1 : 2.1

Molybdenum is an element with a lower degree of toxicity compared to Co, Ni, Cr and is a β-stabilizing element. Field studies [13-16] have highlighted that titanium alloyed with titanium in varying percentages between 15-20% can reduce the modulus of elasticity leading to adequate mechanical properties.

Zirconium is used in medical applications because of low modulus of elasticity, high corrosion resistance, and high biocompatibility with human tissue [6, 29].

Tantalum is considered biocompatible, being a β-stabilizing element that influences the value of the elastic modulus [6, 27].

Titanium is a non-toxic element, even in larger quantities; some studies have shown the influence of ingestion by humans of up to 0.8 mg of titanium daily, proving that titanium was eliminated without being digested / assimilated. Its uses in medical applications were due to good interaction with the host bone, the titanium implants not being rejected by the body and having a high corrosion resistance [27-31].

IV.2 Vacuum Arc Remelting

In order to obtain the TMZT alloys, the MRF ABJ 900 Vacuum Arc Remelting has been used. Advantages of using this equipment are as follows:

- very high melting temperatures can be achieved;
- the possibility of melting the metallic vacuum samples under a protective atmosphere by means of a non-consumable mobile electrode of thorium tungsten;
- creates alloys with uniform composition, through repeated remeltings;
- possibility of mixing elements with different melting temperatures;
- can be used various crucibles for elaboration and ensures the possibility of obtaining under specific conditions the samples in the form of a pill of different shapes and sizes;
- loading and unloading is done in a simple way by lifting the cover that is caught in the hinge to the rest of the camera;
- it is illuminated with a halogen lamp, thus helping to control the melting of the alloying elements in the process [32].

Materials Research Forum LLC

https://doi.org/10.21741/9781644900796

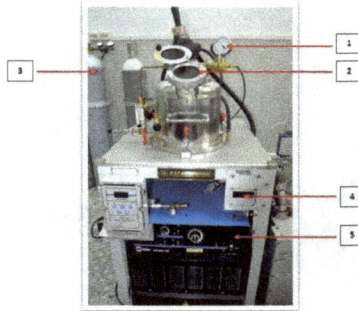

Figure IV.5. Vacuum Arc Remelting Plant MRF ABJ 900: 1-flowmeter for argon pressure control in the furnace; 2-furnace window viewing window; 3- argon bottles; 4-pressure measuring equipment; 5-controlpanelcommand [32].

The technical parameters for the vacuum arc melting furnace (Figure IV.5) are as follows:

- melting power - min. 55 kVA;
- melting current - min. 650 A, 60% DS, three-phase voltage;
- maximum temperature - 3.700°C;
- possibility of remelting Zr, Ti, Co-Cr, Ni-Cr alloys;
- continuous monitoring of the vacuum level with display of its value;
- maximum vacuum obtained with preliminary vacuum and diffusion pumps: 10-6 mbar;
- inert gas supply system - argon;
- the oven chamber is made of 306L stainless steel and the double walls are water-cooled;
- water-cooled copper base plate 230mm (diameter) x 13mm (thickness);
- inconsumable electrode from Tungsten with 6.5mm (diameter) [5].

The furnace can use different copper plates, for the production of samples and according to the dimensions of the samples we want to elaborate, the model of the crucibles chosen. Figure IV.6 shows types of copper plates that can be used to melt metallic loads.

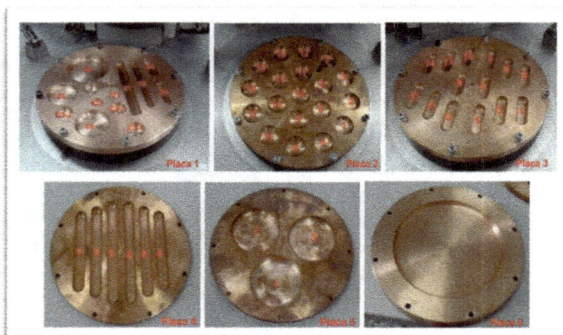

Figure IV.6. Crucibles used for melting alloys for various applications.

Since the cost of titanium alloy casting facilities are high, the method most commonly used by researchers is the melting and casting technique, a fast and affordable technology for small and medium-sized laboratories.

IV.3 The technological flow to obtain TMZT alloys

The study of the influence of the alloying elements on the process of synthesis of the TMZT alloys, the differences between the melting temperatures and the densities of the elements in the alloy composition, the strong chemical reactivity of these elements and their interaction in the liquid state with the gases in the atmosphere of the forming aggregate, they imposed to be melted in vacuum or in an inert controlled atmosphere to ensure complete melting, homogeneity of the distribution of the components in the mass of the alloy and a low level of impurities.

The metallic load for obtaining alloys must be of high quality and purity, degreased and properly prepared. For the preparation of TMZT alloys, high purity elements such as Ti-99.8%, Mo-99.7%, Zr-99.2% and Ta-99.5% were used as starting materials.

Figure IV.7 shows the copper plate pattern chosen to obtain TMZT alloys. The chosen crucible is provided with circular and elongated cavities and is suitable for the amount of material that we want to obtain for the characterization of the properties of the alloys.

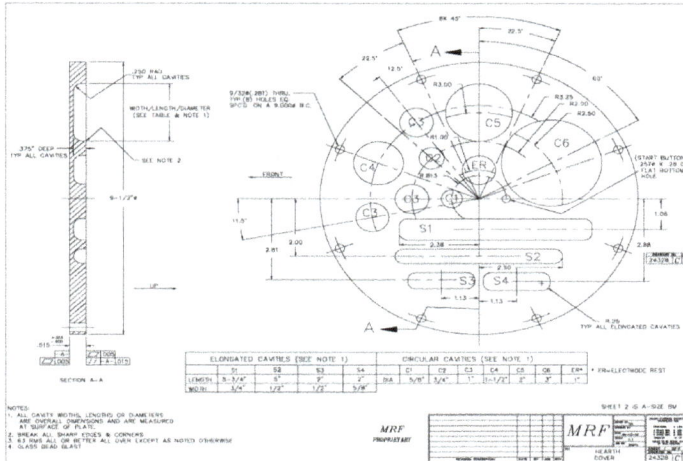

Figure IV.7. Crucible scheme used for obtaining TMZT alloys [10].

The elaboration of the TMZT alloys in the vacuum arc melting furnace comprised a succession of operations as described in Figure IV.8.

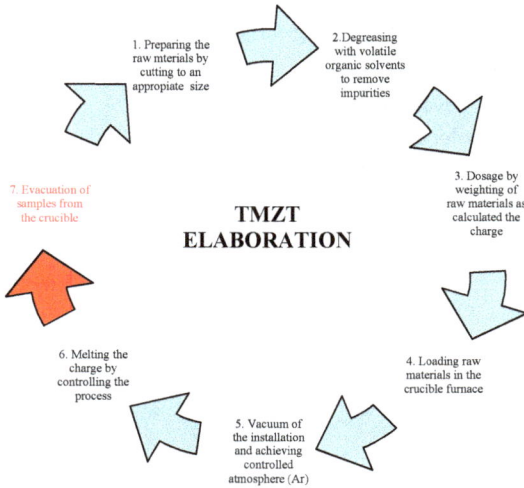

Figure IV.8. Technological flow of Ti-Mo-Zr-Ta alloys.

Figure IV.9 shows all stages of TMZT alloying, which includes the weighing of the raw material, loading of the alloying elements and the final semi-finished obtained products.

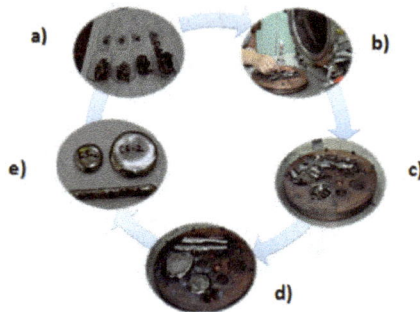

Figure IV.9. Stages of TMZT alloying process: a) weighing of raw materials and gravimetric dosing; b), c) loading of the raw material; d), e) TMZT semi-products obtained after solidification.

The load calculation has taken into account the characteristics of the different alloying elements and their physico-chemical properties. For this, we have to consider the method of designing and constructing the melting furnace used, since they greatly influence the losses of metal upon melting.

Elaboration of the alloys was carried out in three charges to obtain two alloys in each charge. Alloys were developed in the circular cavities C5, C6 and the elongated cavities S1, S2. Table IV.3 shows the cavities used for each alloy.

Table IV.3. Cavities used in the three charges for TMZT alloys.

Alloy	Cavity from crucible used	Number of charge
Ti15Mo7Zr5Ta	C5, S1	1
Ti15Mo7Zr10Ta	C6, S2	
Ti15Mo7Zr15Ta	C5, S1	2
Ti20Mo7Zr5Ta	C6, S2	
Ti20Mo7Zr10Ta	C5, S1	3
Ti20Mo7Zr15Ta	C6, S2	

The raw materials were prepared for melting and dosing for each charge by weighing with an electronic balance, according to the charge calculation.

Table IV.4. Experimental results for TMZT alloys.

Alloy	Cavity from crucible	Ti [g]	Mo [g]	Zr [g]	Ta [g]	Initial charge introduced into the plant [g]	Final charge removed from the plant [g]
Ti15Mo7Zr5Ta	C5-1	51.20	10.52	5.00	3.62	70.34	70.34
	S1-1	73.07	14.97	7.03	5.07	100.14	99.40
Ti15Mo7Zr10Ta	C6-1	135.83	30.15	14.30	20.29	200.57	200.70
	S2-1	63.02	15.08	7.04	15.00	70.55	70.55
Ti15Mo7Zr15Ta	C5-2	44.11	10.70	5.07	10.61	70.49	70.20
	S1-2	63.02	15.08	7.04	15.00	100.14	99.98
Ti20Mo7Zr5Ta	C6-2	136.12	40.10	14.13	10.02	200.37	200.50
	S2-2	47.69	14.03	5.05	3.51	70.28	70.39
Ti20Mo7Zr10Ta	C5-3	44.12	14.03	5.03	7.05	70.25	70.24
	S1-3	63.02	20.04	70.01	10.13	100.59	100.50
Ti20Mo7Zr15Ta	C6-3	116.34	40.11	14.17	30.20	200.82	200.60
	S2-3	40064	14.20	5.01	10.68	70.53	70.53
TOTAL						1325.07	1323.93

Table IV.4 shows the load of the raw material used, resulting from the load calculation for the experimental alloys. The raw material load also took into account the proposed alloying reports for TMZT alloys.

The load introduced into the plant as well as the mass of the semi-finished products after solidification were weighed and presented in Table IV.3. During the elaboration process, splashes of melted material from one cavity to another have been leaked due to the process of the arc melting process, therefore some alloys have a higher value at weighing after solidification. The loading of the raw material in the plant was done in the ascending order of their specific density: tantalum, molybdenum, zirconium and titanium, with the tight sealing of the chamber of the processing plant.

During the melting operations, a vacuum atmosphere of 4.5×10^{-3} mbar was carried out, followed by the purge of the inert gas chamber (Ar), repeated three times to remove the air from the working chamber of the plant.

The development process of the TMZT alloys was controlled and guided throughout the elaboration through the furnace's observation window, a suitably illuminated enclosure.

Elaboration of the TMZT alloys made with a vacuum arc melting system, took place by the melting of the elements, followed by the remeltings of alloys for six times, a necessary operation for the refining and homogenization of the alloys. The melting of the elements took place uniformly resulting alloys with a precise and homogeneous chemical composition. After the solidification, two semifinished samples of each alloy were obtained in the form of ingots, shapes and different masses, but with sufficient quantity for taking the specimens required for all proposed laboratory tests.

For the development of TMZT alloys, the physical, chemical and technological properties, as well as the mutual influences between the alloying elements used, were taken into account. Thus, six alloy variants with Ti, Mo, Zr and Ta alloying elements were developed with the aid of a vacuum remeltings plant. The elaborated alloys were: Ti15Mo7Zr5Ta, Ti15Mo7Zr10Ta, Ti15Mo7Zr15Ta, Ti20Mo7Zr5Ta, Ti20Mo7Zr10Ta, Ti20Mo7Zr10Ta and Ti20Mo7Zr15Ta.

The charge calculation was performed according to the proposed alloying percentages, resulting in the metal load which also took into account the vaporization phenomenon characteristic of the technology.

Two semi-finished products from each alloy were obtained, sufficient to collect for samples required for structural, thermal, surface, mechanical, electrochemical and cytotoxicity laboratory tests. Testing of the specimens who aimed at obtaining standard samples with minimal material losses was carried out by electroerosion.

References

[1]   M.S. Cercel (cas. Bălțatu), Contribuții privind îmbunătățirea proprietăților aliajelor de Ti-Mo destinate aplicațiilor medicale – teză de doctorat, Iași, 2017.

[2]   Information on www.chalcogen.ro

[3]   ***ASM Handbook, Alloy Phase Diagrams, vol.3.

[4]   *** ASM Handbook, Properties and Selected Nonferrous Alloys and Special-Purpose Materials, Vol. 2.

[5]   *** ASM Handbook, Metallography and Microstructure, vol.9.

[6]   A.V. Sandu, M.S. Baltatu, Nabialek M., Savin A., Vizureanu P., Characterization and Mechanical Proprieties of New TiMo Alloys Used for Medical Applications, *Materials* 2019, *12*(18), 2973. https://doi.org/10.3390/ma12182973

[7]   Information on http://www.supraalloys.com

[8]   Information on Handbook of Ternary Alloy Phase Diagrams

[9]   A.C. Bărbînță, Îmbunătățirea proprietăților aliajelor de Ti-Nb-Zr-Ta utilizate la fabricarea protezelor ortopedice - teză de doctorat, Iași, 2003.

[10] V. Geantă, I. Tripșa, R. Ștefănoiu, Tratat de Știința si Ingineria Materialelor, Vol. 2.,

Bazele teoretice şi ingineria obţinerii materialelor metalice, Cap. 7 – Elaborarea oţelurilor. Academia de Ştiinţe Tehnice din Romania, Editura AGIR, Bucureşti, 2007.

[11] N. Dumitraşcu, Biomateriale şi biocompatibilitate, Ed. Universităţii "Alexandru Ioan Cuza", Iaşi, 2007.

[12] D.P. Balaban, Biomateriale, Ed. Ovidius University Press, Constanţa, 2005.

[13] C. Mihăilescu, A. Mihăilescu, N. Georgescu, Biomateriale de uz chirurgical, Ed. Tehnopress, Iaşi, 2011.

[14] C.N. Cumpătă, Implaturi acoperite chimic cu hidroxiapatita biologică, Ed. Printech, Bucureşti, 2012.

[15] C. Lipşa, D.S. Lipşa, Biomateriale-Curs pentruanul I, Iaşi, 2009.

[16] D.C. Ludwigson, Requirements for metallic surgical implants and prosthetic devices. Metals Engineering Quarterly: American Society of Metallurgists 1, 1965.

[17] C. Demian, Cercetări privind comportarea materialelor destinate implantării osoase conform normelor europene de calitate - teză de doctorat, Universitatea "Politehnica", Timişoara, 2007.

[18] R. Chelariu, G. Bujoreanu, C. Roman, Materiale metalice biocompatibile cu baza titan, Ed. Politehnium, Iaşi, 2006.

[19] I. Antoniac, Biomateriale metalice utilizate la executia componentelor endoprotezelor totale de sold, Ed. Printech, Bucureşti, 2007.

[20] D.M. Bombac, M. Brojan, P. Fajfar, F. Kosel, R. Turk, Review of materials in medical applications, Materials and Geoenvironment, vol.54(4), 2007, pp. 471-499.

[21] C.N. Elias, J.H.C. Lima, R. Valiev, M.A. Meyers, Biomedical applications of titanium and its alloys, Biological Materials Science, 2008, pp. 46- 49. https://doi.org/10.1007/s11837-008-0031-1

[22] M. Niinomi, Mechanical properties of biomedical titanium alloys, Mater. Sci. Eng., A, 243 (1998) 231-236. https://doi.org/10.1016/S0921-5093(97)00806-X

[23] L.I. Linkow, Prefabicated mandibular prostheses for intraosseous implants, J Prosthet Dent., 20(4), 1968, p. 367–375. https://doi.org/10.1016/0022-3913(68)90234-5

[24] L.I. LinkowL.I., The blade vent-a new dimension in endosseous implantology, Dent Concepts, 11 (2), 1968, pp.3-12.

[25] G. Lütjering, J.C. Williams, Titanium-Second Edition, Springer Science + Business Media, Germany, 2000.

[26] M. Geetha, A.K. Singh, R. Asokamani, A.K. Gogia, Ti based biomaterials, the ultimate choice for orthopaedic implants - A review, Mater. Sci., 54 (2009) 397-425. https://doi.org/10.1016/j.pmatsci.2008.06.004

[27] D. Bunea, A. Nocivin, Materiale biocompatibile, Ed. şi Atelierele Tipografice Bren, Bucureşti, 1998.

[28] H. Vermeşan, Cercetări privind comportarea la coroziune a otelurilor inoxidabile supuse deformarii plastice si nitrurarii ionice - teza de doctorat, Universitatea Tehnică din Cluj Napoca, 1998.

[29] C. Popa, V. Cândea, V. Şimon, D. Lucaciu, O. Rotaru, Ştiinţa biomaterialelor, Ed. U.T. Press, Cluj-Napoca, 2008.

[30] M.G. Minciună, Contribuţii privind îmbunătăţirea proprietăţilor aliajelor de cobalt utilizate în aplicaţii medicale-Teză de doctorat, Iaşi, 2014.

[31] Q. Chen, G.A. Thouas, Metallic implant biomaterials, Materials Science and Engineering R, 87 (2015) 1–57. https://doi.org/10.1016/j.mser.2014.10.001

[32] Information on www.eramet.ro

CHAPTER V: Investigation of Ti-Alloys Regarding Structural, Mechanical, Electrochemical Characterization and Biocompatibility Assessment

Biomaterials must be tolerated by the body for a long time (decades) and therefore must meet the functional requirements according to the medical applications in which they are to be used.

In this sense, these biomaterials must be investigated from all points of view (chemical, structural, mechanical, biocompatibility), as they will fulfill their best functions and not produce adverse reactions [1-7].

V.1. Structural and Thermal Characterization of Experimental Alloys Obtained

V.1.1 Determination of the Chemical Composition of TMZT Alloys by EDAX Analysis

In order to achieve the structural and thermal characterization it is necessary to know / identify the chemical composition of the alloys obtained. EDAX microanalysis with energy dispersion of X-radiation was used to determine the chemical composition of the Ti-Mo alloys developed. Determination of the chemical composition by EDAX microanalysis is the first laboratory investigation required to highlight the proportions obtained between the pure chemical elements and was performed on TMZT alloys obtained by melting in arc vacuum furnace.

In order to obtain a more accurate determination, five EDX determinations were made in different areas of the samples to check the concentrations obtained.

To determine the chemical composition of the TMZT alloys obtained, the Vega Tescan LMH II equipment was used using the Bruker EDAX attached to the SEM equipment.

For the determination of the chemical composition of alloys obtained from the TMZT system, samples having dimensions of 10 mm x 10 mm x 5 mm were used. Before being examined, the samples were ground on abrasive paper to remove impurities and titanium oxide film on the surface of the alloy.

Table V.1 shows the average of the chemical compositions of the experimental alloys.

Table V.1. Chemical compositions of TMZT alloys, expressed as a mass percentage.

Alloy	Titanium [%]	Molibdenum [%]	Zirconium [%]	Tantalum [%]
Ti15Mo7Zr5Ta	75.46	13.56	6.34	4.64
Ti15Mo7Zr10Ta	73.63	12.33	6.80	7.24
Ti15Mo7Zr15Ta	66.00	12.56	6.99	14.45
Ti20Mo7Zr5Ta	70.89	17.49	6.45	5.17
Ti20Mo7Zr10Ta	60.35	21.68	7.17	10.80
Ti20Mo7Zr15Ta	63.36	18.45	6.48	11.71

Table V.1 shows the mass percentages of the elements identified in the alloy composition, the percentages of the elements varying slightly with the theoretical batch calculation. The analysis bulletins on the chemical composition obtained revealed that the main elements identified in the alloys elaborated are: Ti, Mo, Zr, Ta, without the presence of alloy inclusions.

In Figure V.1 an EDX spectrum characteristic of TMZT alloys is highlighted.

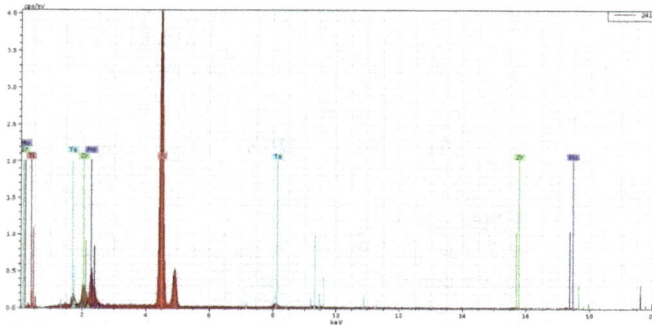

Figure V.1. EDX spectrum for TMZT alloys.

X-ray microanalysis with X-ray energy dispersion (EDAX) determined the chemical composition of the alloys elaborated. This analysis confirmed the presence of the chemical elements Ti, Mo, Zr, Ta in the composition of the analyzed alloys, respectively the mass percentages of these elements, which are consistent with the theoretical calculations regarding the elaboration process. Also, no chemical inhomogeneities resulting from the rapid solidification process have been identified, which are frequently found in the case of Ti-Mo alloys.

V.1.2 Microstructural Characterization of TMZT Alloys by Optical Microscopy

Microscopic methods of structural analysis are used to characterize materials based on their structure, constituents and phases present (nature, shape, dimensions and distribution) and possible structural defects (pores, cracks, structural inhomogeneity's, etc.) [8]. Structure analysis was performed using the OPTIKA XDS-3 MET microscope.

In order to investigate the metallographic structure, the preparation of the metallographic samples of the experimental alloys TMZT included a sequence of steps: cutting to appropriate dimensions (eg. 10 mm x 10 mm x 5 mm), incorporation in epoxy resin, grinding and polishing at specific speeds, chemical attack with specific reagents (10 mL of HF, 5 mL of HNO_3, 85 mL of H_2O for 30 s). After the steps were taken, the samples were analyzed at the optical microscope at various magnification powers to obtain detailed images on the microstructure.

Figures V.2-V.7 highlight images obtained by optical microscopy for TMZT alloys at 50x and 100x magnification.

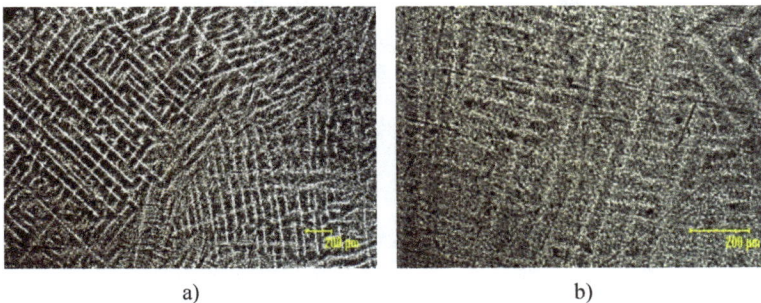

a)   b)

Figure V.2. The optical microstructure of the Ti15Mo7Zr5Ta alloys investigated at a magnification power of: a) 50x, b) 100x.

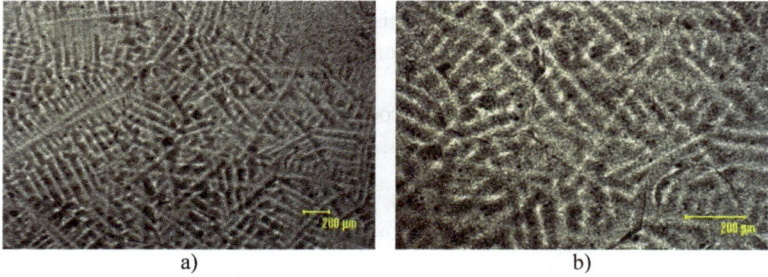

a)   b)

Figure V.3. The optical microstructure of the Ti15Mo7Zr10Ta alloys investigated at a magnification power of: a) 50x, b) 100x.

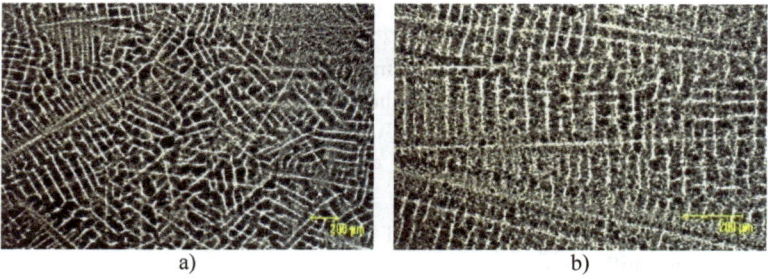

a)   b)

Figure V.4. The optical microstructure of the Ti15Mo7Zr15Ta alloys investigated at a magnification power of: a) 50x, b) 100x.

a)   b)

Figure V.5. The optical microstructure of the Ti20Mo7Zr5Ta alloys investigated at a magnification power of: a) 50x, b) 100x.

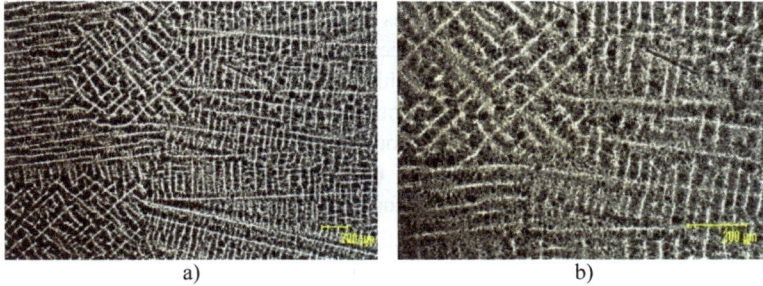

Figure V.6. The optical microstructure of the Ti20Mo7Zr10Ta alloys investigated at a magnification power of: a) 50x, b) 100x.

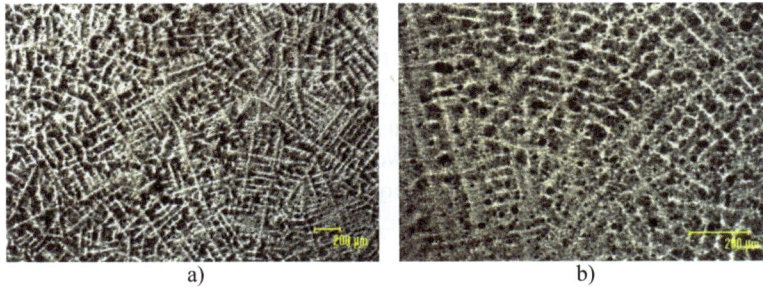

Figure V.7. The optical microstructure of the Ti20Mo7Zr15Ta alloys investigated at a magnification power of: a) 50x, b) 100x.

Figures V.2-V.7 show the structure of TMZT alloys with titanium alloy grain specifics. Images obtained by optical microscopy for Ti20Mo7Zr5Ta, Ti20Mo7Zr10Ta, Ti20Mo7Zr15Ta have a dendritic structure with irregular grain boundaries. These acicular and coarse structures are specific to β alloys. The Ti15Mo7Zr5Ta, Ti15Mo7Zr10Ta, and Ti15Mo7Zr15T alloys clearly show the lamellar dendrites inside the β -grains.

The optical microstructure of TMZT alloys has been shown to have acicular structures with irregular dendritic grain boundary limits (Ti20Mo7Zr5Ta, Ti20Mo7Zr10Ta and Ti20Mo7Zr15Ta), as well as large grain structures of type β with dendrite inside (Ti15Mo7Z5Ta, Ti15Mo7Zr10Ta and Ti15Mo7Zr15Ta), distributed uniforms.

The variation of the α, α + β and β phases consists in the differences in the chemical composition of the constituents. The high percentage of β (Mo, Ta) stabilizing elements led to the formation of a β-type structure, highlighted very well in TMZT alloys. Also, zirconium in concentrations below 10% contributes to the refining of the microstructure, thus allowing the formation of a homogeneous and uniformly distributed structure. Thus, elements with percentages of tantalum (5-15%), combined with a molybdenum concentration of 15-20%, contribute to the formation of β phase [1-7].

Optical microscopy reveals a uniform biphasic structure, consisting of a high proportion of solid β solution, in which the dendritic lamellar structures of the orthorhombic martensite α" are present. Orthorhombic martensite occurs frequently in the case of titanium alloys containing β-stabilizers of the transition metals category, including molybdenum and tantalum. In the present case, the presence of the α" phase is due to the decomposition of the β phase during the cooling [1-7].

V.1.3 Structural Characterization of TMZT Alloys by X-ray Diffraction

X-ray diffraction is a method of non-destructive analysis, being used in the study of materials for the determination of the crystal structure by measurements on the symmetry and dimensions of the crystalline grid, as well as by determining the place occupied by atoms in the elemental cell. X-ray diffraction is used to study the imperfections of the crystalline grid, the magnitude and domain form of coherent scattering and distortions within them [9].

Phase establishing was performed by qualitative X-ray diffraction analysis using a Panatical X'Pert Pro MPD equipment. Thus, the phases and compounds that make up the investigated alloys are highlighted. Parameters used for sample analysis are: an angle θ-2θ range between 20° -80°; continuous type scanning; step size of 0.0131303 (°), time per step: 60 (s); scanning speed 0.054710 (o/s); number of steps: 6093. An X-ray tube with copper anode was used which emits X-rays in linear mode using a Pixcel detector.

The processing of the obtained data was done with the Highscore Plus program, than they were imported and processed using experimental data processing software in order to obtain diffractograms of the experimental alloys.

Figures V.8-10 shows the diffractograms for the investigated TMZT alloys.

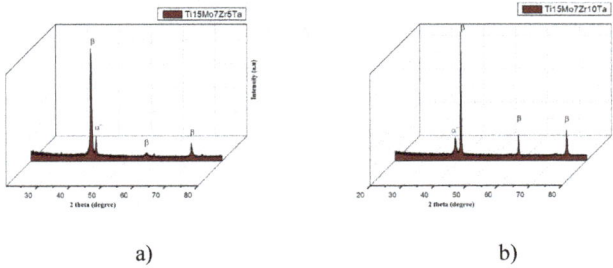

a)   b)

*Figure V.8. Diffractograms of experimental alloys: a) Ti15Mo7Zr5Ta,
b) Ti15Mo7Zr10Ta.*

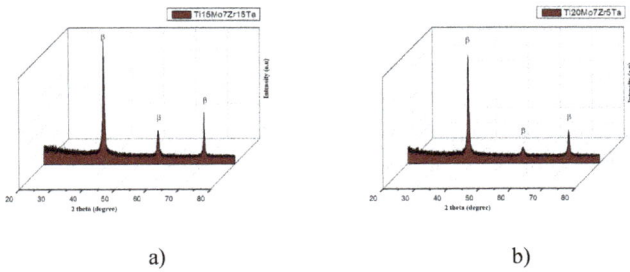

a)   b)

*Figure V.9. Diffractograms of experimental alloys: a) Ti15Mo7Zr15Ta,
b) Ti20Mo7Zr5Ta.*

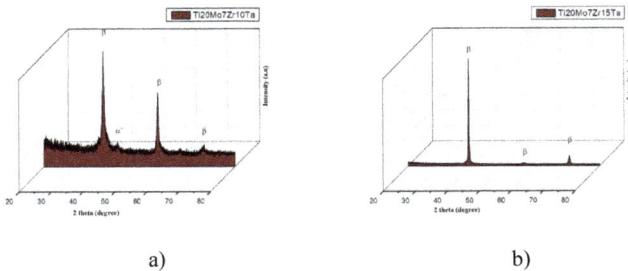

a)   b)

*Figure V.10. Diffractograms of experimental alloys: a) Ti20Mo7Zr10Ta,
b) Ti20Mo7Zr15Ta.*

In order to determine the constituent phases, by X-ray diffractometry, 10 mm x 10 mm x 5 mm samples of TMZT alloys were used after they had been polished.

X-ray diffraction revealed a biphasic structure composed of two main phases: β with centered cubic structure and α "with orthorhombic structure.

Diffractograms resulting from the investigation of TMZ alloys consist of a succession of diffraction peaks, with the order of X diffraction radiation measured in pulses / second, and the abscissa angle γ, where θ is the Bragg angle measured in normal degrees. With the aid of diffractograms the structural constituents can be determined.

The diffractograms discussed above confirm the β-type structures identified through optical microscopy, taking into account that titanium is an allotropic element, being presented and observed in different forms: up to a temperature of 882°C, having a α-Ti compact hexagonal structure and above 882°C, β-Ti, having a centered cube structure. In the composition of the investigated alloys there is a major phase β with a centered cube structure and a secondary phase α" with an orthorhombic structure.

Phase β, having a centered cubic structure, is fixed by alloying the titanium with transition elements (Mo, Ta), while phase α" (HC) is a martensite, crystallized in the orthorhombic system and is formed when the content of stabilizing elements phase b, decomposed during cooling [1-7, 10, 11].

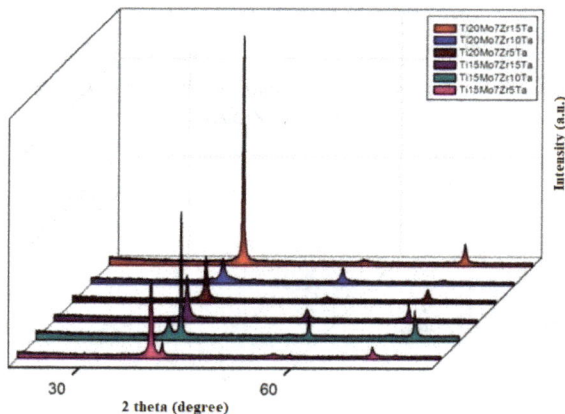

Figure V.11. Comparative Diffractograms of Experimental Alloys.

The predominant phase β (Figure V.11) for the investigated samples: Ti15Mo7Zr5Ta, Ti15Mo7Zr10Ta, Ti15Mo7Zr15Ta, Ti20Mo7Zr5Ta, Ti20Mo7Zr10Ta, Ti20Mo7Zr15Ta, Ti20Mo7Zr15Ta, was identified with the main peak at the angle 2θ = 29.4669°; 40.3030°; 38.2958°; 58.7960°; 37.2459°; 38.9277°.

Parameters of compounds, such as crystallographic system, network parameters, or cell volume are highlighted in Table V.2.

Table V.2. The crystallographic parameters for the phases identified following the XRD analysis.

Solid solution	Space group	System of crystallization	a (Å)	b (Å)	c (Å)	α (°)	ß (°)	γ (°)	Volume cell (10^6 pm3)	RIR
TaTi	Fd-3m	Cubic	3.29	3.29	3.29	90	90	90	35.48	19.73
MoTi	Fd-3m	Cubic	3.17	3.17	3.17	90	90	90	31.86	14.59

From Table V.2 it can be seen that after the volume analysis of the TMZT alloys, two solid solutions have been identified having the cubic crystallization system.

V.1.4 Thermal Scanning by Differential Scanning Calorimetry (DSC)

To determine the thermal behavior of the elaborated TMZT alloys, a thermal analysis method was used in which the temperature of the sample to be studied and that of the reference sample was linear and the difference in heat quantity was measured along with any variation in mass.

For the thermal analysis by differential calorimetric analysis of TMZT alloys, samples were cut at sizes corresponding to the mass up to 50 mg, the dimensions being imposed by the dimensions of the crucibles in which the samples were tested. From the measurements made after sampling, it was found that the effective mass was in a range of 42-49 mg.

To characterize thermal behavior, the target temperature range is between 36.5-37.2°C, within which a biomaterial works inside the healthy human body.

The thermograms recorded following the DSC analysis performed with the PROTEUS program are shown in Figures V.12-14.

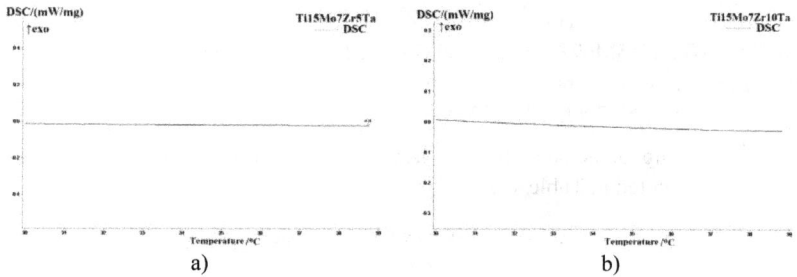

Figure V.12. Thermograms obtained for experimental alloys: a) Ti15Mo7Zr5Ta, b) Ti15Mo7Zr10Ta.

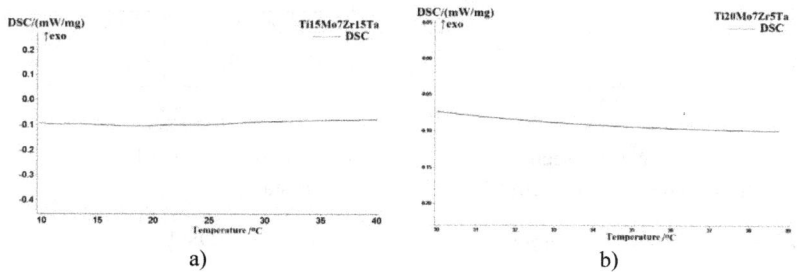

Figure V.13. Thermograms obtained for experimental alloys: a) Ti15Mo7Zr15Ta, b) Ti20Mo7Zr5Ta.

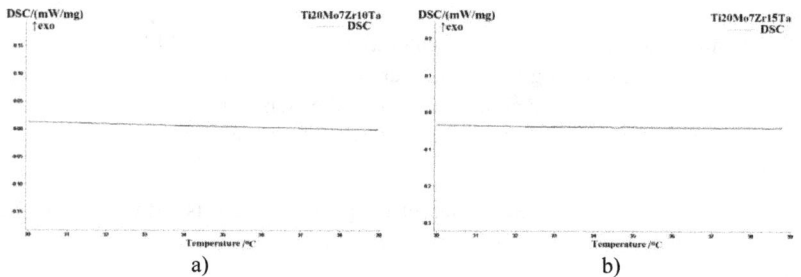

Figure V.14. Thermograms obtained for experimental alloys: a) Ti20Mo7Zr10Ta, b) Ti20Mo7Zr15Ta.

The thermal regime to which the analyzed samples were subjected was:

a. heating from room temperature to 40°C at a rate of 10 K / min;

b. cooling at 10 K / min up to room temperature.

The thermograms of investigated TMZT alloys did not record phase transformations in the temperature range 30-40°C. As a result of the thermal analysis, the investigated alloys show stable values in the temperature range of 36.5-37.2°C, according to the literature, where the phase change of the titanium alloys takes place over 882°C.

Following thermal analysis by DSC, TMZT alloys showed thermal stability at the temperature of the human body, showing good interaction with the cellular environment for use as medical implants.

Thermal analysis by differential scanning calorimetry revealed that TMZT alloys did not register phase transformations in the range of 30-40°C, being stable at the temperature of the human body [6, 12].

V.2. Mechanical Properties Characterization of the TMZT Alloys

Knowledge of the mechanical properties of the TMZT, was performed to determine the characteristics of hardness, the indentation test, tensile test, respectively, compression testing and fractographic analysis.

V.2.1 Hardness Characteristics Determination of TMZT Alloys

The hardness measurements highlight the resistance to the penetration action of an external body and provide information on the behavior of the studied materials. In this way, we can analyze the TMZT alloys developed for the purpose of fitting them into the specific medical application.

The hardness measurements made on the TMZT alloys developed by the Vickers method are a general method for determining the hardness of metallic materials, widely used in biomaterial testing [1-7].

The hardness measurements were made on samples of TMZT alloys of dimensions 10 mm x 10 mm x 3 mm, the surface of the samples being prepared by grinding on abrasive paper. The experimental tests consisted of three determinations in different areas on the surface of each sample, using a 9.81 N pressure force and a 12 second measurement time.

In Table V.3.the results obtained from the hardness measurements of investigated TMZT alloys are presented, and in Figure V.15. is the graphical representation of TMZT alloys.

Table V.3. The hardness values of TMZT alloys measured by the Vickers method.

Alloy	Ti15Mo 7Zr5Ta	Ti15Mo 7Zr10Ta	Ti15Mo 7Zr15Ta	Ti20Mo 7Zr5Ta	Ti20Mo 7Zr10Ta	Ti20Mo 7Zr15Ta
Average value (HV)	379.21	462.33	390.88	388.86	321.31	397.56

From the results obtained from the hardness tests, the Ti20Mo7Zr10Ta (321.31 HV) alloy showed the lowest value and the highest hardness value was recorded by the Ti15Mo7Zr10Ta alloy.

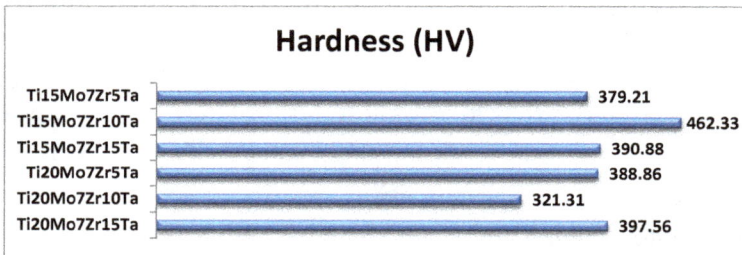

Figure V.15. Graphic representation of hardness for the hardness values of TMZT alloys measured by the Vickers method.

According to the literature [10, 12] and the results obtained from the experimental research on hardness on the TMZT alloys, a comparative study was made between the lowest and the highest value of HV hardness of the TMZT alloys and the classical biomaterials made of stainless steel, the Ti6Al4V alloy, and the CoCrMo class of alloys (Figure V.16).

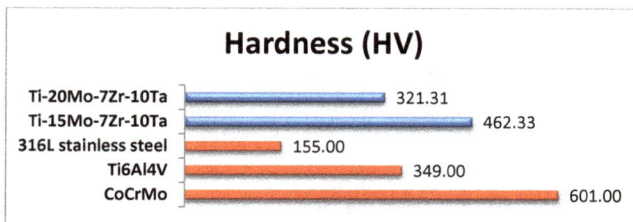

Figure V.16. Graphical comparison for the hardness values of TMZT alloys with other biomaterials.

Compared to other biomaterials, TMZT alloys have a higher hardness than stainless steels, but close to the Ti6Al4V alloy, which is most commonly used in implantology [13, 14].

V.2.2. Some of Mechanical Properties Characterization of TMZT Alloys by Indentation Method

Indentation is a test method based on the principles used to determine the modulus of elasticity, rigidity, etc. The indentation tests were performed using Universal Micro-Tribometer CETR UMT-2 tribological and mechanical assay equipment.

Samples of TMZT alloys were cut with dimensions of 17 mm x 5 mm x 5 mm and the surface of the samples was prepared by grinding with high-grain silicon carbide abrasive paper and polished with alumina suspension to obtain metallic shine and removal surface roughness. The investigated samples were fixed on a flat surface of the test device by means of screws and clamps. The tests were carried out under dry conditions. A Rockwell diamond-type penetrator was used having an angle at the 120° incisor cone and a 200 μm spherical sphere tip, applying a force of 5 N.

Three determinations for each alloy were made to determine as accurately as possible. After the workflows and their recording by the UMT 2 software, the imprinting curves (depth vs. force) of the TMZT alloys were plotted using the VIEWER program [1-7].

Figures V.17-19 shows the response of alloys during indentation tests in the form of force-depth dependencies.

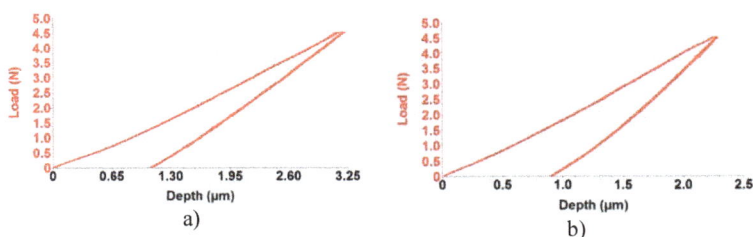

Figure V.17. The strength-depth curve of the microindentance test for investigated TMZT alloys: a) Ti15Mo7Zr5Ta, b) Ti15Mo7Zr10Ta.

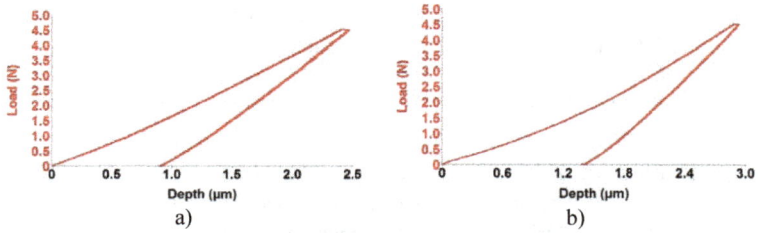

Figure V.18. The strength-depth curve of the microindentance test for investigated TMZT alloys: a) Ti15Mo7Zr15Ta, b) Ti20Mo7Zr5Ta.

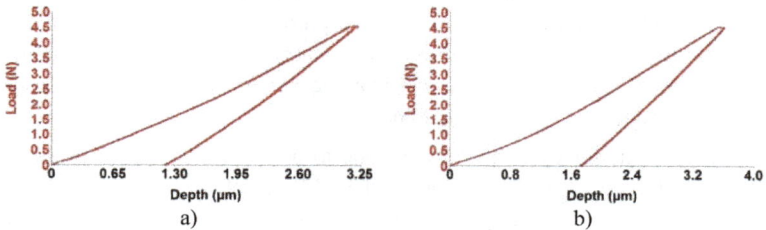

Figure V.19. The strength-depth curve of the microindentance test for investigated TMZT alloys: a) Ti20Mo7Zr10Ta, b) Ti20Mo7Zr15Ta.

The values of the elasticity modulus of TMZT alloys resulting from the indentation test are highlighted in Table V.4 and plotted in Figure V.20.

Table V.4. Elasticity young modulus values for TMZT alloys measured by the indentation test.

Alloy	Ti15Mo 7Zr5Ta	Ti15Mo 7Zr10Ta	Ti15Mo 7Zr15Ta	Ti20Mo 7Zr5Ta	Ti20Mo 7Zr10Ta	Ti20Mo 7Zr15Ta
Young modulus (GPa)	69.02	76.88	51.93	51.68	43.41	43.57

The elastic modulus is a very important criterion to the choice of metallic materials used in orthopedics, and should be as close as possible to human bone (17-30 GPa) [15-17].

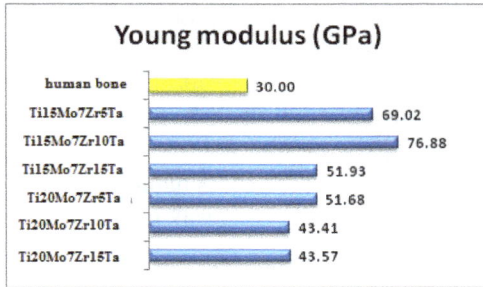

Figure V.20. Graphic representation of TMZT alloy modulus values measured by the indentation test.

The investigated TMZT alloys have values between 43.41 to 76.88 GPa for the modulus of elasticity, measured by indentation tests. The lowest value is the Ti20Mo7Zr10Ta (43.41 GPa) alloy, and the highest value is Ti15Mo7Zr10Ta (76.88 GPa). The low elasticity modulus of the investigated alloys is due to the presence of β-stabilizing elements such as Mo and Ta [18-22]. According to Figure V.20, it can be observed that with the increase of the Mo content from 15% to 20% and with the increase of the percentage of Ta from 10% to 15%, they lead to the decrease of the modulus of elasticity by approximately 25 GPa .

Increasing the Mo ratio by 5% and maintaining the percentage of Zr and Ta constantly contributed to a 44% decrease in the modulus of elasticity for TMZT alloys.

The comparison of values obtained for TMZT alloys with those of classical biomaterials is presented in Figure V.21.

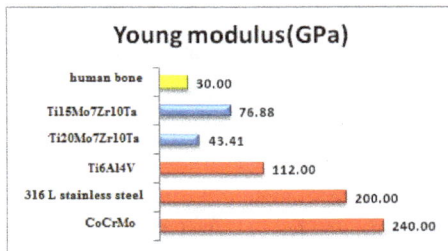

Figure V.21. Graphical comparison of TMZT alloy modulus values with other metallic biomaterials.

Compared with other metallic biomaterials: CoCrMo alloys (210-253 GPa) and stainless steels (190-210 GPa), the investigated TMZT alloys exhibit much lower values and the closest values to the human bone (17-30 GPa). TMZT alloys show a significant improvement in mechanical properties even from titanium alloys: Ti6Al4V (100-114 GPa) and C.P. Ti (102-104 GPa).

From the point of view of elasticity, TMZT alloys fulfill the criteria imposed on metallic biomaterials, the values of the modulus of elasticity being close to those of the human bone [1-7].

V.2.3 Determination of Mechanical Properties by Tensile Strength Test

The types of requests found in the materials used in medical applications are very varied, so it is necessary to determine all the mechanical characteristics. The basic test of the materials is the tensile strength test is to determine the mechanical strength.

In biomedical applications, external stresses acting on various body elements are generally composed, inducing both normal stresses σ and tangential stresses τ. Based on these, it is calculated by considering one of the resistance theories, an equivalent tensile of the type:

$$\sigma_{ech} = \frac{\sigma}{2} + \frac{1}{2}\sqrt{\sigma^2 + 4\tau^2} \leq \sigma_a \tag{5.1}$$

where σ_a is the admissible tensile that can be determined based on one of the following expressions:

$$\sigma_a = \frac{\sigma_r}{3}; \sigma_a = \frac{\sigma_c}{2} \tag{5.2}$$

where σ_r represents the tensile strength and σ_c represents the flow limit.

From TMZT alloys, only four samples were cut to standard sizes due to the insufficient quantity of material that was subjected to the tensile strength test. The main geometric parameters for investigated TMZT samples are illustrated in Figure V.22.

The equipment used for the tensile strength test was a universal test machine INSTRON 8801 with hydraulic sample gripping jets. The test was performed at ambient temperature with a loading velocity of 0.5 N / min, allowing taking into consideration as many essential points on the characteristic curve $\sigma - \varepsilon$ (stress-strain curve).

Figure V.22. Shape and dimensions of rectangular section samples used for traction applications.

The samples were tested by elongation along its main axis at a constant speed until breakage. Characteristic curves obtained from the traction test of TMZT alloys are shown in Figures V.23-24 [1-7].

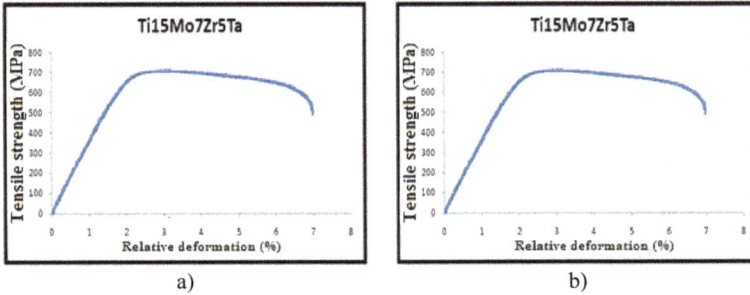

a)

b)

Figure V.23. Characteristic curves and results obtained from the tensile strength test of samples from the investigated TMZT alloys: a) Ti15Mo7Zr5Ta, b) Ti15Mo7Zr15Ta.

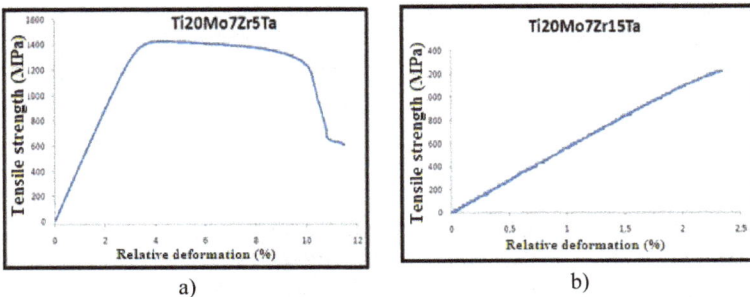

a)

b)

Figure V.24. Characteristic curves and results obtained from the tensile strength test of samples from the investigated TMZT alloys: a) Ti20Mo7Zr5Ta, b) Ti20Mo7Zr15Ta.

From the characteristic curves recorded by the TMZT alloys, the behavior of the alloys can be observed at the tensile strength request. TMZT alloys exhibited a different behavior until breakage, some being ductile, some fragile. The ductile had a tear elongation of 0.07-0.10 mm / mm while the fragile sample had a tensile elongation of 0.02 mm / mm (Ti10Mo7Zr15Ta).

The results of the experimental determinations obtained by the tensile strength test of the elaborated TMZT alloys can be found in Table V.5.

Through the traction test of TMZT alloys, the following mechanical characteristics were determined:

- Tensile Strength (MPa);
- Flow Limit (MPa);
- Elongation at break (mm);
- Modulus of elasticity (MPa);
- Energy to Break (J);
- Elongation (%);
- Strangled (%).

Table V.5. Mechanical characteristics measured by tensile strength test for the investigated TMZT alloys.

Alloy	Tensile Strength (MPa)	Flow Limit (MPa)	Elongation at break (mm)	Modulus of elasticity (MPa)	Energy to Break (J)	Elongation (%)	Strangled (%)
Ti15Mo7Zr5Ta	712.29	636.10	0.07	36390.63	2.35	4.75	22.22
Ti15Mo7Zr15Ta	1051.74	969.90	0.07	40123.67	2.97	2.50	30.55
Ti20Mo7Zr5Ta	1410.38	1272.63	0.10	51311.36	6.76	4.75	30.55
Ti20Mo7Zr15Ta	1223.21	1100.00	0.02	56918.20	0.78	1.25	16.66

The graphical representation of the tensile resistance values of the TMZT alloys measured by the tensile strength test is shown in Figure V.25. Tensile strength resistance values for TMZT alloys ranged from 712.29 MPa (Ti15Mo7Zr5Ta) to 1410.38 MPa (Ti20Mo7Zr5Ta). Thus, an average traction resistance of 1099.41 MPa was obtained. It can be seen that as the percentage of molybdenum increases, the tensile strength increases. Compared to Co-Cr alloys (900-1000 MPa) and stainless steels (500-1350 MPa) [14, 18], TMZT alloys (Ti15Mo7Zr15Ta, Ti20Mo7Zr15Ta, Ti20Mo7Zr5Ta) show higher values, which proves that the alloys have mechanical properties very good, except the Ti15Mo7Zr5Ta alloy.

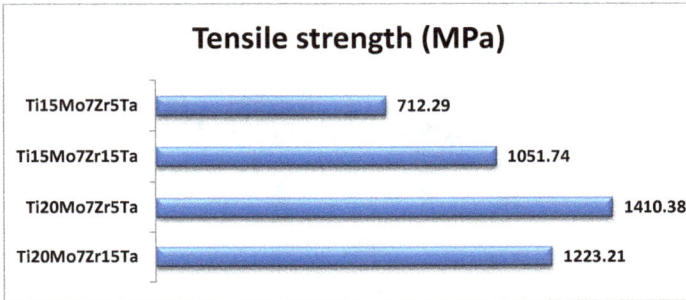

Figure V.25. Graphic representation of traction resistance values of TMZT alloys measured by tensile strength test.

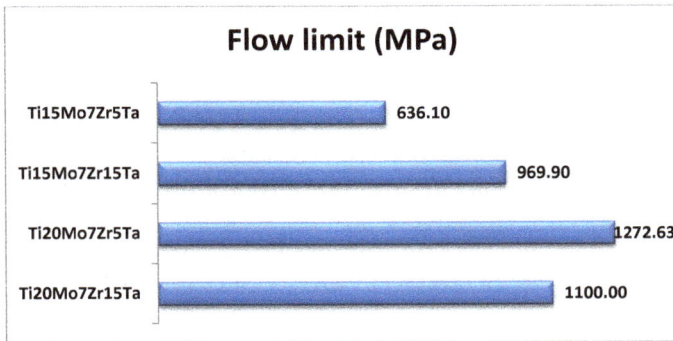

Figure V.26. Graphic representation of TMZT alloy flow limit values measured by tensile strength test.

The flow limit values of the TMZT alloys measured by the tensile strength test were between of 636.10 - 1272.63 MPa.

Considering three of the mechanical characteristics, like the tensile strength, the flow limit and the total energy accumulated in the sample before the break, it is found that Ti15Mo7Zr5Ta shows the best behavior.

The graphical representation of values for the young modulus of elasticity for the TMZT alloys measured by the tensile strength test is shown in Figure V.27.

The flow limit values as well as the tensile strength of the TMZT alloys show the positive influence of molybdenum with increasing concentration and the constant concentration of zirconium and tantalum.

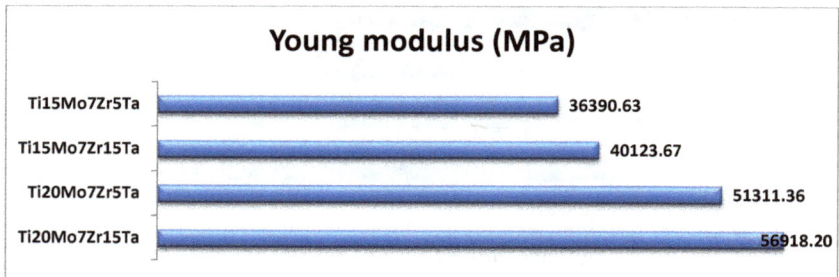

Young modulus (MPa)

Alloy	Value
Ti15Mo7Zr5Ta	36390.63
Ti15Mo7Zr15Ta	40123.67
Ti20Mo7Zr5Ta	51311.36
Ti20Mo7Zr15Ta	56918.20

Figure V.27. Graphic representation of TMZT alloy young modulus values measured by traction test.

The longitudinal elastic modulus (E), also called Young's modulus, is defined as the ratio between normal stress σ and specific deformation ε, and is the property of a material to deform and return to its original form. Biomaterials should have a modulus of elasticity as close as possible to that of the structure of the system into which they are introduced, respectively to the human bone.

The values of the elasticity modulus of TMZT alloys measured by the tensile strength test were in the range 36390.63 - 56918.20 MPa and 36.39-56.92 GPa respectively. The lowest value of the module was the Ti15Mo7Zr5Ta alloy, and the highest value was the Ti20Mo7Zr15Ta alloy. With regard to the results of TMZT alloys compared to other biomaterials, these alloys have the closest value to human bone, which recommends their use in medical applications.

The indentation method and the tensile strength test have provided different results but with approximately similar values of the modulus of elasticity. This is because the determination of the modulus of elasticity by the compression indentation method at a 5N load was done on micron-sized surface areas, whereas the modulus of elasticity determined by the traction test is related to the entire area of the sample section.

Another mechanical characteristic is the response of the impact sample, estimated by the value of the absorbed breaking energy. The change in potential impact energy (between pre-impact and post-impact value) is highlighted on a calibrated dial that shows the total energy absorbed during sample breakage. The energy up to breakage of TMZT alloys varied between 0.78 J (Ti20Mo7Zr15Ta) and 6.76 J (Ti20Mo7Zr5Ta).

Based on the resulting values of the mechanical properties of TMZT alloys, equal or even better values were obtained compared to classical biomaterials, recommending them to be used successfully in medical applications.

V.2.4 Determination of Mechanical Properties by Compression Test

Compressive strength is a very important mechanical feature for biomaterials for making suitable implantable materials (rods or joints, etc.) because it provides information on how to behave in operation (complex tensile strength and compression loads, corrosion and fatigue).

For the compression stress test, the samples from the experimental alloys were processed by milling (from the melted samples) and then cut to the nominal dimensions of 5 mm x 5 mm x 5 mm using a metalographic abrasive disc (thickness 1, 25 mm) and coolant. From measurements made after cutting, it was found that the effective sample sizes had values ranging from 5.5 to 4.60 mm. To obtain accurate data, two samples were tested for each type of alloy.

Samples of TMZT alloys subjected to the compression test did not show non-metallic inclusions or other observable defects. The compression test machine was Walter + Bai AG and the test was carried out at ambient temperature, with mechanical characteristics such as: flow limit, compressive strength, specific shortening and buckling resistance.

The initial dimensions of the samples (L1,2i, [mm], initial height H0 [mm]) were measured before the compression test, then after testing, obtaining the final width values L1,2f [mm] and the heights final Hu [mm] (Figure V.28).

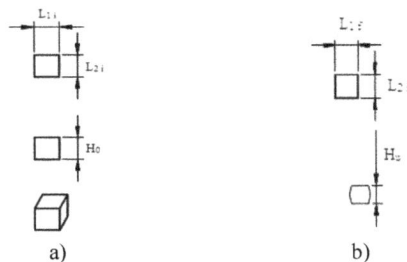

Figure V.28. Measured dimensions of the samples a) before, b) after the test.

Deformation of samples from TMZT alloys was performed in accordance with the minimum resistance law, by reducing the height and rounding of the settlement base, with evolution to the "barrel" shape (Figures V.29-31).

a)   b)

Figure V.29. Appearance of TMZT alloy samples after compression test:
a) Ti15Mo7Zr5Ta, b) Ti15Mo7Zr10Ta.

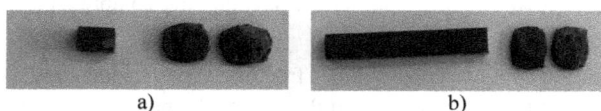

a)   b)

Figure V.30. Appearance of TMZT alloy samples after compression test:
a) Ti15Mo7Zr15Ta, b) Ti20Mo7Zr5Ta.

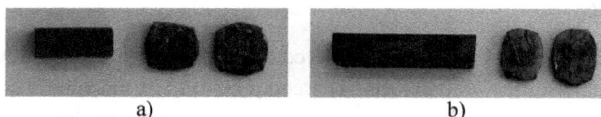

a)   b)

Figure V.31. Appearance of TMZT alloy samples after compression test:
a) Ti20Mo7Zr10Ta, b) Ti20Mo7Zr15Ta.

The values of the geometric characteristics of the samples subjected to the compression test (two determinations for each sample) are presented in Table V.6, the meaning of the symbols being the following: H0,u[mm] - initial / final height; L1,2 i, f [mm] - initial / final sample sizes; S0, u [mm^2] - the initial / final cross section of the samples.

The following were calculated from the initial and final measured values: the initial / final cross-section of the sample (S0,u mm^2), sample reduction (ΔH%) and barrel (deformation) of the sample (Zc %). The calculation relationships used are shown in Table V.7.

Table V.6. Values of samples measured before and after the compression test of TMZT alloys.

Sample[*]	H_0 [mm]	L_{1i} [mm]	L_{2i} [mm]	S_0 [mm^2]	H_u [mm]	L_{1f} [mm]	L_{2f} [mm]	S_u [mm^2]
Ti15Mo7Zr5Ta-1	5.05	4.97	4.63	23.01	2.72	6.77	7.24	49.01
Ti15Mo7Zr5Ta-2	4.60	4.55	5.03	22.89	1.86	8.13	8.67	70.49
Ti15Mo7Zr10Ta-1	4.88	4.93	5.24	25.83	1.86	9.52	8.97	85.39
Ti15Mo7Zr10Ta-2	4.86	5.22	4.94	25.79	2.18	8.58	8.07	69.24
Ti15Mo7Zr15Ta-1	5.32	4.92	4.75	23.37	1.94	8.63	9.04	78.02
Ti15Mo7Zr15Ta-2	5.25	4.95	4.58	22.67	2.10	8.47	8.31	70.39
Ti20Mo7Zr5Ta-1	4.91	5.48	5.04	27.62	1.94	9.00	9.51	85.59
Ti20Mo7Zr5Ta-2	4.91	5.03	5.07	25.50	1.81	9.11	9.23	84.09
Ti20Mo7Zr10Ta-1	5.15	5.36	4.93	26.42	2.20	9.06	8.45	76.56
Ti20Mo7Zr10Ta-2	5.13	4.98	5.35	26.64	2.09	9.26	8.81	81.58
Ti20Mo7Zr15Ta-1	4.98	4.98	4.89	24.35	1.86	8.97	9.13	81.90
Ti20Mo7Zr15Ta-2	5.54	4.91	4.96	24.35	1.71	9.96	9.94	99.00

*)Note: Two samples were tested for each alloy

Table V.7. Calculation relationships for determining the characteristics of TMZT alloys by the compression test [19, 20].

Typical mechanical characteristics	Calculation relationship
Shortening the sample	$A_c = \dfrac{(h_0 - h_u)}{h_0} \cdot 100 \ [\%]$
Barrel sample	$Z_c = \dfrac{(S_u - S_0)}{S_0} \cdot 100 \ [\%]$
Flow limit	$\sigma_c = \dfrac{F_c}{S_0} \ [\text{N/mm}^2]$

Values obtained by calculation for ΔH - sample shortening and Zc - sample barrel are shown in Table V.8.

Table V.8. Values of shortening and barrel at the compression test for TMZT experimental alloys.

Sample*	ΔH [%]	Zc [%]
Ti15Mo7Zr5Ta-1	46.14	53.05
Ti15Mo7Zr5Ta-2	59.57	67.53
Ti15Mo7Zr10Ta-1	61.89	69.75
Ti15Mo7Zr10Ta-2	55.14	62.76
Ti15Mo7Zr15Ta-1	63.53	70.04
Ti15Mo7Zr15Ta-2	60.00	67.79
Ti20Mo7Zr5Ta-1	60.49	67.73
Ti20Mo7Zr5Ta-2	63.14	69.67
Ti20Mo7Zr10Ta-1	57.28	65.48
Ti20Mo7Zr10Ta-2	59.26	67.34
Ti20Mo7Zr15Ta-1	62.65	70.26
Ti20Mo7Zr15Ta-2	69.13	75.40

)Note: Two samples were tested for each alloy

Curves showing the evolution of stress / strain characteristics resulting from the compression test for TMZT alloys are shown in Figures V.32 - V.37.

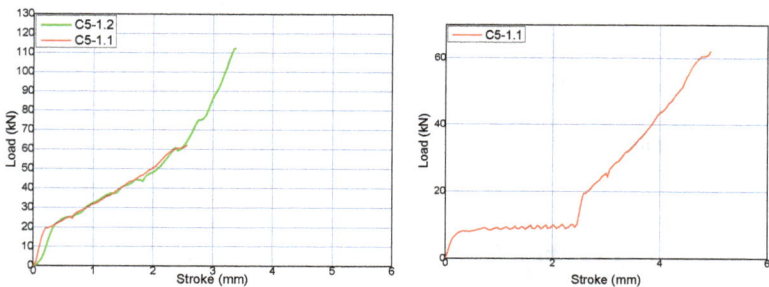

Figure V.32. The curves resulting from the compression test for the Ti15Mo7Zr5Ta alloy. Note: Since C5-1.1 did not have parallel surfaces due after cutting, the curve was corrected. To the right is the uncorrected curve.

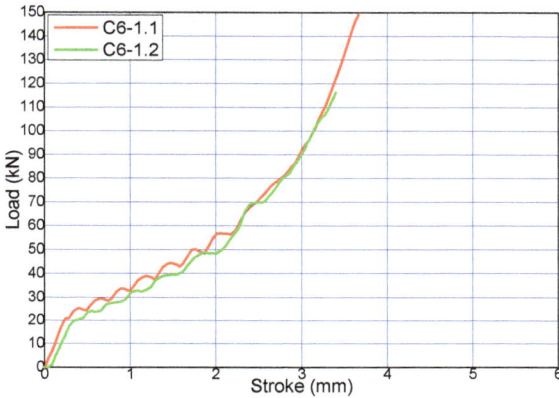

Figure V.33. The curve resulting from the compression test for the Ti15Mo7Zr10Ta alloy.

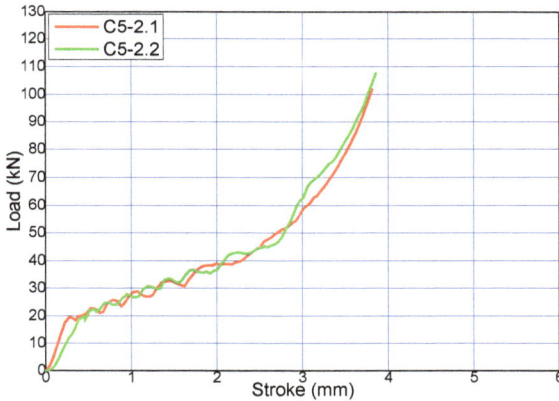

Figure V.34. The curve resulting from the compression test for the Ti15Mo7Zr15Ta alloy.

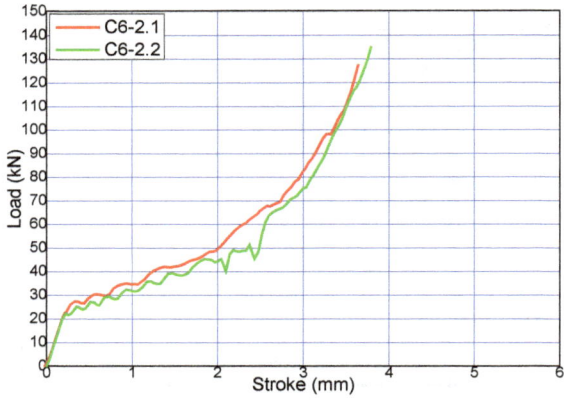

Figure V.35. The curve resulting from the compression test for the Ti20Mo7Zr5Ta alloy.

Figure V.36. The curve resulting from the compression test for the Ti20Mo7Zr10Ta alloy.

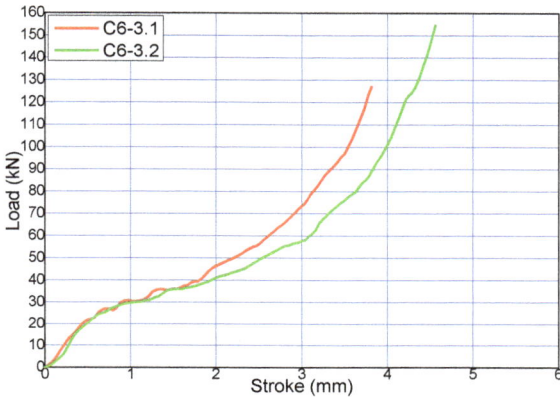

Figure V.37. The curve resulting from the compression test for the Ti20Mo7Zr15Ta alloy.

From the analysis of the compression deformation diagrams of the TMZT alloys (Figures V.32-V.37) it appears that the studied materials are tenacious, deflecting plastic continuously as the load increases. In none of the samples there have not been identified any cracks on the surfaces of the forming.

From the data presented in Table V.6 it is noted that of the percentage barrels are relatively close to the all analyzed alloys (ranging from 62.76 - 75.40%) except the Ti15Mo7Zr5Ta-1 sample, where the percentage of the barell sample was lower 53.05%), which can be attributed to the fact that there have been inclusions that have led to a decrease in the value of the barrels.

Because tantalum has a high compressibility and low cytotoxicity, this chemical element has been added to Ti-based alloys to make it easier to process by severe plastic deformation. Analyzing the TMZT alloy binder values, a direct proportional increase of this parameter is observed with the increase in tantalum concentration, with values between 67.53% (Ti15Mo7Zr5Ta for sample 2) to 70.04% ((Ti20Mo7Zr10Ta for sample 1) for one 15% molybdenum, and 65.48% (Ti20Mo7Zr10Ta for sample 1), respectively, to 75.40% (Ti20Mo7Zr15Ta for sample 2) with 20% Mo.

For higher values of molybdenum concentration (20% Mo) and variable tantalum concentrations, it is noted that as the percentage of tantalum increases, the material brute is stronger, proving that the material is more easily deformable and has good machinability by deformation cold plastic. The cumulative effects of alloying with 20%

molybdenum and a percentage rate between 5-15% tantalum lead to 3-5% higher barrels compared to alloyed 15% molybdenum alloys.

For the compression strength estimation, the compression and flow limit (σc) values in the compression test (Table V.9) must be analyzed.

Table V.9. Deformation force (Fp) values and compression flow limit (σc) of experimental TMZT alloys.

Sample*	Fp [kN]	σc [MPa]
Ti15Mo7Zr5Ta-1	16.89	734.04
Ti15Mo7Zr5Ta-2	17.44	762.20
Ti15Mo7Zr10Ta-1	15.80	611.77
Ti15Mo7Zr10Ta-2	15.84	614.44
Ti15Mo7Zr15Ta-1	13.10	560.46
Ti15Mo7Zr15Ta-2	11.78	519.69
Ti20Mo7Zr5Ta-1	19.98	723.41
Ti20Mo7Zr5Ta-2	19.89	779.94
Ti20Mo7Zr10Ta-1	12.85	486.32
Ti20Mo7Zr10Ta-2	13.33	500.28
Ti20Mo7Zr15Ta-1	14.69	603.23
Ti20Mo7Zr15Ta-2	13.93	571.99

\)Note: Two samples were tested for each alloy

The fact is noticed that at both 15% Mo and 20% Mo concentrations, the flow limit values are lower (the material is softer) as the percentage of tantalum increases. Some unevenness (hardening tendency) is found in the 20% Mo alloy, which registered a significant increase in the flow limit of 486.32 MPa for the Ti20Mo7Zr10Ta alloy to 603.23 MPa for the Ti20Mo7Zr15Ta alloy. Compressive stiffness is shown in Figure V.38, cumulative for all samples analyzed. This figure shows the high deformability of the analyzed samples, which exhibits evolutions on all deformation layers (P1-P2), until the zone of instability (P3).

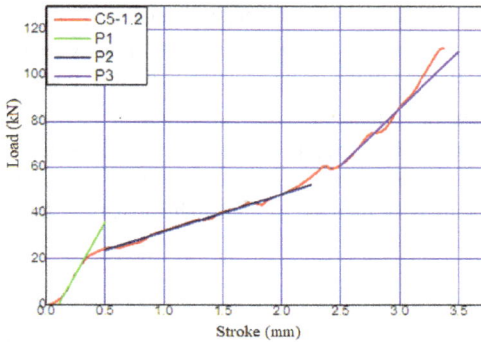

Figure V.38. Estimation of compression stiffness by analyzing the slope of the force - displacement curves, where: P1 - the linear region of the curve - elasto - plastic deformation; P2 - material flow (uniform strain deformations, P3 - non-uniform deformations – instability.

Table V.10. Synthesis of deformation mode on different stages of the force-position curve for experimental alloys.

Sample*	P1 [kN/mm]	P2 [kN/mm]	P3 [kN/mm]
Ti15Mo7Zr5Ta-1	118.53	17.72	-
Ti15Mo7Zr5Ta-2	90.75	16.31	49.93
Ti15Mo7Zr10Ta-1	90.06	18.95	80.94
Ti15Mo7Zr10Ta-2	80.03	18.34	62.31
Ti15Mo7Zr15Ta-1	93.61	10.60	58.97
Ti15Mo7Zr15Ta-2	62.36	11.62	65.90
Ti20Mo7Zr5Ta-1	114.69	13.91	92.43
Ti20Mo7Zr5Ta-2	115.94	13.44	77.23
Ti20Mo7Zr10Ta-1	47.72	15.79	52.48
Ti20Mo7Zr10Ta-2	51.66	14.28	59.70
Ti20Mo7Zr15Ta-1	49.07	11.56	62.62
Ti20Mo7Zr15Ta-2	57.69	10.57	78.33

) Note: Two samples were tested for each alloy

In conclusion, it can be said that the analyzed experimental alloys have satisfactory values of the compressive strength characteristics, predominant being the tendency for deformation in the initial elastic phase (P1) and in the final unstable plastic flow phase (P3).

Since no imperfections or a tendency of fragmentation have been identified in the compressed samples, it can be said that the alloys analyzed have good deformability under severe conditions at ambient temperature. The effect of alloying with tantalum and molybdenum is beneficial as an increase in the batch capacity of the samples subjected to compression has been achieved.

V.2.5 Fractographic Analysis

Breaking surface analysis allows the evaluation of the behavior of different mechanical stresses applied to the investigated alloys. Aspects of breakage surfaces of samples from TMZT experimental alloys resulting from tensile testing were analyzed by scanning electron microscopy (Inspect S, SEM) to highlight breakage at various magnifications.

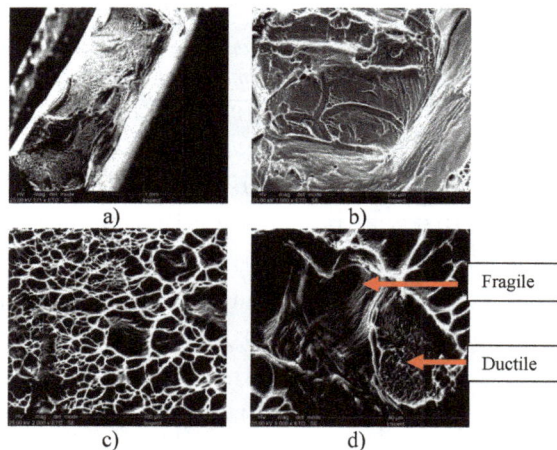

Figure V.39. Aspect of the surface of the sample analyzed from the Ti15Mo7Zr5Ta alloy at different magnification powers: a) 121X; b) 1000X; c) 2000X; d) 5000X.

In Figure V.39 a), b), c), d), we can see the appearance of the Ti15Mo7Zr5Ta alloy surfaces. Surface analysis in different areas highlights a mixed rupture, containing ductile and fragile dots, Figure V.39 d). Details are shown with a fragile fracture area (typical appearance of cleavage - plain sliding plane - Figure V.39 b) and a ductile fracture area (typical appearance with recesses and connecting mesh - Figure V.39 c), areas the ductile breaking properties being preponderant.

Figure V.40 shows the appearance of the breaking surfaces for the Ti15Mo7Zr15Ta alloy. The macroscopic images show mixed fracture zones, Figure V.40 a), c), d) and a predominantly brittle fracture detail b).

The appearance of the Ti20Mo7Zr5Ta alloy breaking faces is shown in Figure V.41. Mixed rupture areas a), b) and tensile areas at breaking are visible. Figure V.41 c) highlights a breakage crevice in a predominantly fragile area surrounded by ductile fracture areas and in Figure V.41 d) a detail is observed on a ductile fracture area and compounds of different density (area white) [1-7].

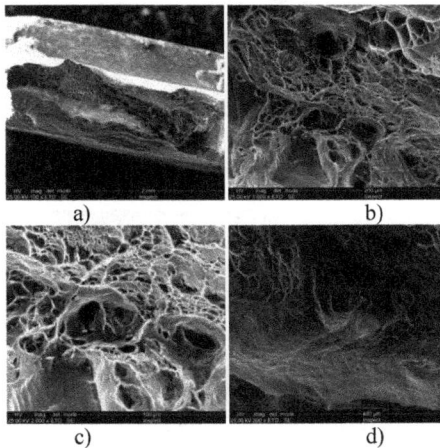

a)   b)

c)   d)

Figure V.40. Aspect of the surface of the sample analyzed from the Ti15Mo7Zr15Ta alloy at different magnification powers: a) 100X; b) 1000X; c) 2000X; d) 5000X.

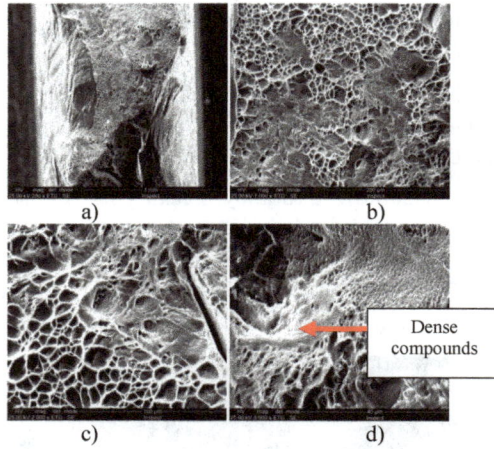

Figure V.41. Aspect of the surface of the sample analyzed from the Ti20Mo7Zr5Ta alloy at different magnification powers: a) 200X; b) 1000X; c) 2000X; d) 5000X.

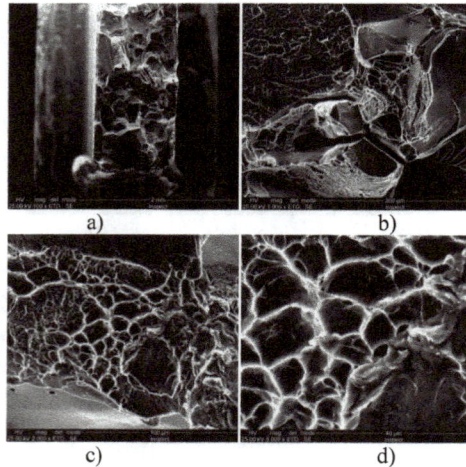

Figure V.42. Aspect of the surface of the sample analyzed from the Ti20Mo7Zr15Ta alloy at different magnification powers: a) 100X; b) 1000X; c) 2000X; d) 5000X.

The appearance of the breaking faces for the Ti20Mo7Zr15Ta alloy is highlighted in Figure V.42. Figures V.42 a) and b) reveal mixed fracture areas (ductile and fragile), the fragile fracture areas being predominant and the appearance of breakage being intergranular. Some details on a micro-ductile fracture area are shown in Figures V.42 (c) and (d) [1-7].

In conclusion, the experimental alloys analyzed as "TMZT" are characterized by mixed breakage (ductile and fragile), some of which are predominantly ductile rupture (Ti20Mo7Zr5Ta) and in others fragile rupture (Ti20Mo7Zr15Ta). The appearance of rupture reveals a tendency for the formation of removing the material before tearing, type crevices, bounded by ductile fracture zones (Ti20Mo7Zr5Ta), which indicates a low homogeneity of the alloy that generated the premature rupture at forces less than those provided for the alloying class.

The physico-mechanical properties of biomaterials, in addition to other biocompatibility characteristics, are crucial in selecting a biomaterial for use in medical applications. The way in which materials can be processed to produce finished parts is defined by their technological properties.

Testing of materials through laboratory investigations was in line with international standards and using analytical methods and test procedures specific to the type of alloy.

Hardness HV measurements showed values in the 321.31 HV ranges (Ti20Mo7Zr10Ta) and 462.33 HV (Ti15Mo7Zr10Ta). The hardness measurements revealed the resistance of the TMZT alloys, recording approximate values of the Ti6Al4V alloy.

Indentation tests have highlighted the values of the elastic modulus. Values of TMZT alloys ranged from 43.57 to 76.88 GPa. Compared with other biomaterials, TMZT alloys show the values closest to those of the human bone (17-30 GPa). These low elasticity values are due to the presence of stabilizing β elements such as Mo, Zr and Ta in titanium. It has been found that by introducing molybdenum in a concentration of 20% versus 15% and 10% tantalum, the modulus of the elasticity modulus is improved by 44%.

The traction test revealed mechanical properties such as: tensile strength, elongation, flow limit and modulus of elasticity. The traction tests revealed that by constantly maintaining the percentage of zirconium and tantalum, an increase in breaking strength was observed with the increase in the percentage of molybdenum.

TMZT alloys have satisfactory compressive strength characteristics, exhibiting good deformability under severe conditions at ambient temperature. The effect of alloying with tantalum and molybdenum is beneficial as an increase in the batch capacity of the samples subjected to compression has been achieved.

Fractographic analysis of TMZT alloys showed different modes of breakage: mixed (Ti20Mo7Zr5Ta) and fragile (Ti20Mo7Zr15Ta). Fractographic analysis studies using the scanning electron microscope were correlated with traction results.

V.3. Electrochemical Characterization of TMZT Alloys in Simulated Biological Environment

Corrosion is the phenomenon of partial or total destruction of materials, especially metals, following chemical or electrochemical reactions to interaction with the environment or with specific environments, the important parameter that determines the reliability of a biomaterial [1].

When implanting a biomaterial into the body, corrosion can lead to the formation of reaction products or metal ions that can damage the health of the body. Their release from biomaterial can cause allergic reactions and damage to adjacent soft tissues.

The corrosion of a biomaterial is determined by the aggressive corrosive nature of the elements existing in body fluids. In order to determine the corrosion potential of titanium alloys developed, it was necessary, to investigate them by potentially dynamic and potentiostatic tests in simulated biological environments.

The simplest method of measuring the corrosion rate of a metal involves putting it in contact with the test medium and measuring the amount of material lost by the sample depending on the exposure time.

Basic chemical processes that occur during corrosion are based on electron exchange as follows:

• Acceptance of electrons is a reduction process and always occurs at the cathode;

• Loss of electrons is an oxidation process and always takes place at the anode.

By plotting the Tafel curves the following parameters were determined which characterize the corrosion resistance of investigated TMZT samples:

- Corrosion potential (E0 [mV]). The corrosion potential represents the chemical equilibrium potential from which the corrosion process starts (the higher the value of E0 is closer to zero, the more chemically stable and harder to corrode).
- The slope of the cathode curve (βc), the slope of the anode curve (βa). The oxidation / reduction characteristics are highlighted in the Tafel diagram on the two branches: the anode branch (βa [mV]) gives indications of the oxidizing character of the corrosive environment and the cathode branch (βc [mV]) provides information on the reduction character.

- Corrosion current density (Jcor). The flow of electrons through the corrosion cell is a measure of the intensity of the corrosion process and is called corrosion current. For this to truly reflect the intensity of a corrosion process, it is related to the surface unit of the corrosion material and is called the corrosion current density.

- Resistance to polarization (Rp) can be used for quantitative description in comparison to the corrosion resistance of metals in different corrosion environments. By high Rp values of metals is meant increased corrosion resistance, and by low values of Rp is meant a low corrosion resistance.

- Corrosion speed Vcor(μm / year). The corrosion rate is due to the oxidation / reduction reactions occurring at the anode and cathode level.

- The determination of the corrosion potential and the drawing of the cyclical/linear polarization curves were performed with a Volta Lab 21 Economical electrochemical system. The potentiostat has special facilities for accurate determination of polarization resistance, recording of corrosion potential and tests on the type of corrosion (in points, generalized) [1].

In order to obtain the electrochemical parameters characterizing the corrosion resistance of elaborated TMZT alloys, the method of electrochemical study was used by linear and cyclic polarization. These methods can directly and quantitatively determine the corrosion rate.

The electrical component of an electrochemical cell consists of a voltmeter measuring the potential of the product. A reference electrode was a saturated calomel electrode in a KCl solution whose potential is reproducible and has a value of 242 mV at a temperature of 25°C. Also a platinum electrode was used as an auxiliary electrode. The data processing was done with the Volta Master 4 program, then imported and processed using experimental data processing software to obtain Tafel and Cyclic diagrams.

The behavior of TMZT alloys during the electrochemical corrosion test at linear polarization is represented in Figures V.43-45 by Tafel diagrams.

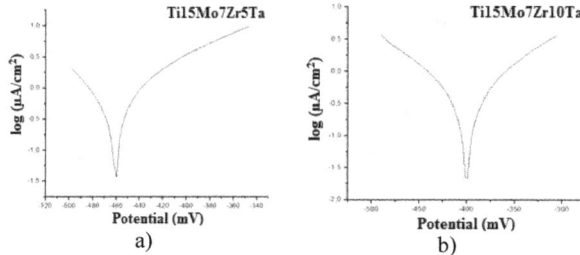

Figure V.43. Tafel diagrams of alloys investigated by corrosion test in Ringer's solution: a) Ti15Mo7Zr5Ta, b) Ti15Mo7Zr10Ta.

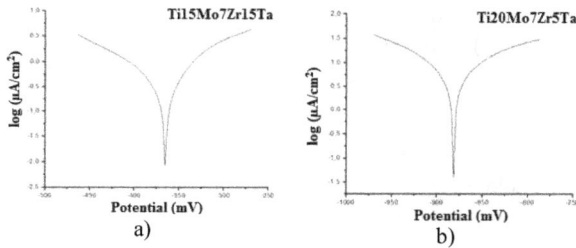

Figure V.44. Tafel diagrams of alloys investigated by corrosion test in Ringer's solution: a) Ti15Mo7Zr15Ta, b) Ti20Mo7Zr5Ta.

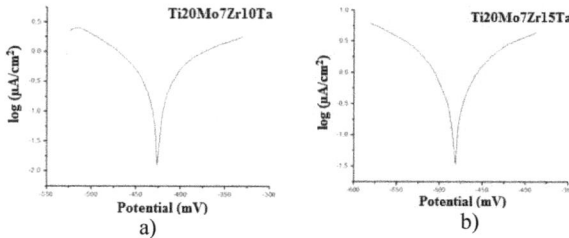

Figure V.45. Tafel diagrams of alloys investigated by corrosion test in Ringer's solution: a) Ti20Mo7Zr10Ta, b) Ti20Mo7Zr15Ta.

In order to determine the corrosion potential of TMZ alloys, samples with dimensions of 5 mm x 5 mm x 5 mm were used. Before being subjected to the corrosion potential determination, the samples were cut and ground to remove impurities and Ti oxide film that formed on the surface of the alloy, then they were embedded in Teflon.

The specific TMZT specimens were introduced into the electrochemical cell and the selected corrosion medium was Ringer's Solution with the composition: NaCl: 8.6 g / L, KCl: 0.3 g / L, CaCl$_2$: 0.48 g / L, for the purpose of to investigate their use as potential biomaterials.

Electrochemical parameters for TMZT alloys immersed in Ringer's solution are shown in Table V.11.

Table V.11. The main parameters of the corrosion process for TMZT alloys in Ringer's solution [1].

Alloy	E_0 [mV]	β_a [mV]	β_c [mV]	R_p [kΩ/cm^2]	J_{cor} [μA/cm^2]	V_{cor} [μm /an]
Ti15Mo7Zr5Ta	-461.20	117.90	-39.50	22.00	1.08	12.67
Ti15Mo7Zr10Ta	-400.10	92.10	-91.20	46.22	0.37	4.31
Ti15Mo7Zr15Ta	-365.40	88.30	-115.00	38.78	0.47	5.49
Ti20Mo7Zr5Ta	-681.60	96.40	-82.10	12.81	0.94	9.28
Ti20Mo7Zr10Ta	-425.50	130.20	-104.30	50.33	0.38	4.47
Ti20Mo7Zr15Ta	-482.00	159.30	-132.80	25.52	1.00	14.05

According to the results obtained in the electrochemical corrosion process for TMZT alloys, the Ti15Mo7Zr10Ta and Ti20Mo7Zr10Ta alloys show close values, with no significant differences.

The corrosion potential (E0) has values close to -400 [mV] for five of the six TMZT samples. The best alloy, which has a long delay corrosion, is Ti15Mo7Zr15Ta having an E0 = -365.4 [mV], suggesting that it has a very high chemical passive surface chemical film stability. The oxidation sensitive corrosion potential alloy is Ti20Mo7Zr5Ta, having E0 = -681.6 [mV] and is due to the occurrence of intergranular chemical segregations.

The βa [mV] oxidation characteristics and the βc [mV] reduction of the six alloys show an increasing trend with increasing the percentage of molybdenum, except for the Ti15Mo7Zr5Ta alloy. The Ti15Mo7Zr15Ta alloy has a ratio of βa / βc = 88.3 / -115.00, net favorably to the reduction characteristics, and the Ti15Mo7Zr10Ta alloy has a balanced ratio of βa / βc = 92.10 / 91.20, these two alloys having corrosion dynamics very low. All other alloys have an imbalance in the βa / βc ratio and a slight oxidation tendency. The sample with the greatest $\beta a / \beta c$ imbalance is Ti15Mo7Zr5Ta ($\beta a / \beta c$ = 117.90 / -39.50).

Corrosion resistance (Rp) is a feature of both the passive outer layer that can recover during corrosion when $\beta c > \beta a$ (predominantly reducing processes to the detriment of oxidants) and also the substrate.

It can be seen from Table V.11 that the very good corrosion resistance samples are: Ti20Mo7Zr10Ta (Rp = 50.33 [kΩ/cm^2]), Ti15Mo7Zr10Ta (Rp = 46.22[kΩ/cm^2]) and Ti15Mo7Zr15Ta (Rp = 38.78 [kΩ/cm^2])

The other samples have a medium corrosion resistance, ranging from 15-25 [kΩ/cm^2], the lowest corrosion resistance test being Ti20Mo7Zr5Ta, which has a corrosion potential (E0) of 25% higher than the other samples.

The corrosion current density Jcor [μA / cm^2] gives indications of the intensity of the corrosion process.

Studying Table V.11 we can see that the lowest corrosion rates have the samples with the ratio $\beta a/\beta c$ favorable to the reduction, balanced or with a slight imbalance (less favorable to oxidation), namely: Ti15Mo7Zr10Ta (Jcor = 0.3681 [μA / cm^2]), Ti15Mo7Zr15Ta (Jcor = 0.47 [μA / cm^2]), Ti20Mo7Zr10Ta (Jcor = 0.38 [μA / cm^2]). The other samples have a Jcor of approximately 1 [μA / cm^2], resulting in a relatively high intensity of the corrosion of the samples.

Corrosion velocity is the intrinsically bounded parameter of the effect of corrosion on the material. It is noted that good corrosion rates recorded Ti15Mo7Zr10Ta (Vcor = 4.3 μm / year), Ti15Mo7Zr15Ta (Vcor = 5.49 μm / year) and Ti20M7Zr10Ta (Vcor = 4.47 μm / year). Samples with medium corrosion rate are Ti15Mo7Zr5Ta (Vcor = 12.67 μm / year) and Ti20Mo7Zr5Ta (Vcor = 9.28μm / year). The only sample whose corrosion rate is three times higher than the most resistant sample is Ti20Mo7Zr15Ta (Vcor = 14.05 μm / year).

The corrosion rate is correlated with both polarization resistance (direct proportional dependence) and corrosion potential (E0).

For all the experimental alloys, the cyclic voltamograms (CV), shown in Figures V.46-48, on the -1,2 ÷ 1,8V range at a scanning velocity v = 10 mV / s were recorded, which revealed generalized corrosion, at the return potential sweep there is a hysteresis characteristic of this type of corrosion.

Generalized corrosion is achieved over a long time given the large reaction area. Corrosion in points (pitting) has a shorter forming period. The variation of the cyclic diagram results in a generalized tendency of corrosion, with very small inflections and modifications of variation.

Titanium is widely used in medical applications due to corrosion resistance. The studies presented, have shown that titanium alloys have a better corrosion resistance than cobalt alloys and stainless steels. Titanium resists water, acids or salt solutions, having a comparable behavior to platinum in the case of chemical corrosion.

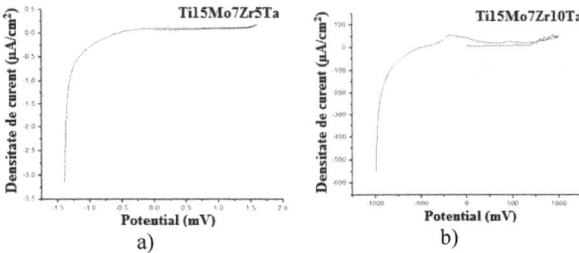

Figure V.46. Cyclical diagrams of alloys investigated by corrosion test in Ringer's solution: a) Ti15Mo7Zr5Ta, b) Ti15Mo7Zr10Ta.

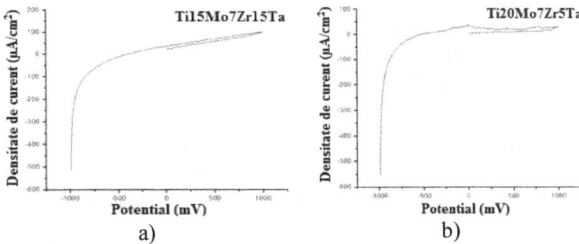

Figure V.47. Cyclical diagrams of alloys investigated by corrosion test in Ringer's solution: a) Ti15Mo7Zr15Ta, b) Ti20Mo7Zr5Ta.

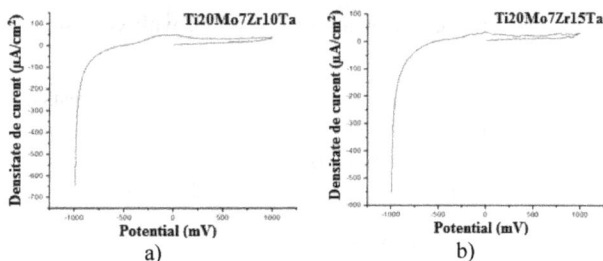

Figure V.48. Cyclical diagrams of alloys investigated by corrosion test in Ringer's solution: a) Ti20Mo7Zr10Ta, b) Ti20Mo7Zr15Ta.

The reason for the corrosion resistance of titanium is the formation, even when exposed to atmospheric oxygen at ambient temperature, of an oxide film that fulfills the three passivation conditions: it is stable, impermeable and very adherent.

In order to confirm and highlight the generalized corrosion type of elaborated TMZT alloys, the surface morphology was highlighted by electronic microscopy (Figures V.49-V.54) at magnification powers of 1200X and 5000X, respectively.

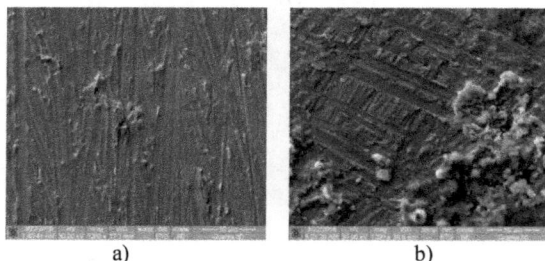

Figure V.49. a), b) Surface area of the sample analyzed after corrosion test for Ti15Mo7Zr5Ta alloy.

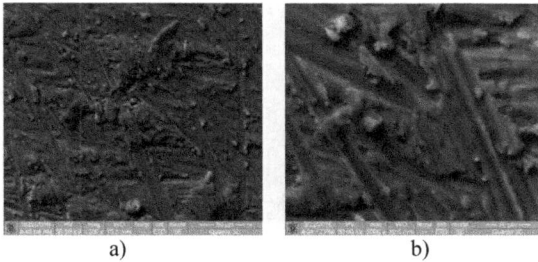

Figure V.50. a) b) Surface area of the sample analyzed after corrosion testing for Ti15Mo7Zr10Ta alloy.

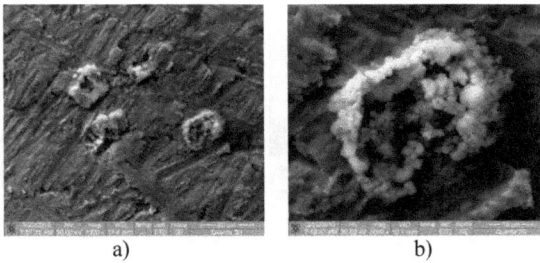

Figure V.51. a), b) Surface area of the sample analyzed after corrosion testing for the Ti15Mo7Zr15Ta alloy.

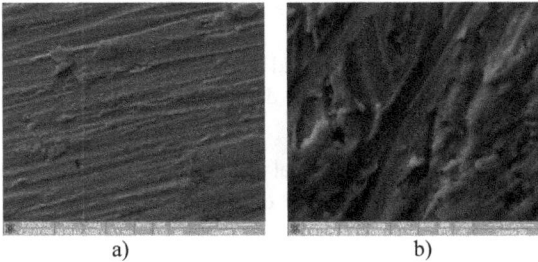

Figure V.52. a), b) Aspect of the sample surface analyzed by corrosion test for Ti20Mo7Zr5Ta alloy.

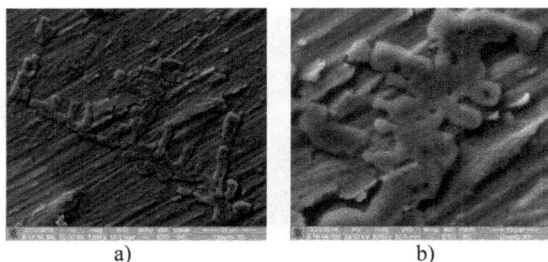

Figure V.53. a), b) Surface area of the sample analyzed after corrosion testing for the Ti20Mo7Zr10Ta alloy.

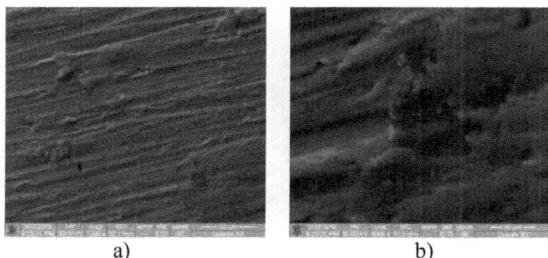

Figure V.54. a), b) Surface area of the sample analyzed after corrosion test for Ti20Mo7Zr15Ta alloy.

After investigating the alloys by linear and cyclic polarization it was found that TMZT alloys elaborated exhibit surfaces little damaged by corrosion, without pitting-like corrosion.

SEM micrographs and surface chemical compositions highlight the formation of chemical compounds in Ringer's solution on the surface of the alloy, respectively Na salts.

Analyzing the surface of the samples subjected to electrocorrosion in simulated biological medium, we observe slight bumps that follow the imperfections of the surface (irregularity due to processing and deepened after corrosion). Also, deposits of existing salts, mainly in the depths, can also be observed. At high magnification powers (5000X) we observe the form of salt deposition in concentric layers.

The electrochemical behavior of TMZT alloys was investigated by the electrochemical study method (linear and cyclic polarization), and the selected corrosion medium was Ringer's solution.

The value of concentrations of molybdenum and tantalum in the alloys studied is important in corrosion studies. The higher the percentage of tantalum, the better the corrosion resistance characteristics can be observed. Molybdenum has a particular influence on corrosion parameters, so for a 15%, corrosion resistance is very good. For 20% Mo alloys there were fluctuations in the corrosion resistance value, recording good corrosion resistance samples corrosion resistance as well as samples with low corrosion resistance.

The results indicate that the passive oxide film formed on TMZT alloys is compact, a characteristic specific to titanium alloys, indicating good corrosion resistance.

V.4. Evaluation of Cellular Viability

V.4.1 Determination of Contact Angle (θ)

Biocompatible materials, such as titanium alloys, have been used and are used in the form of implants that come into contact with living tissues. The wettability of biomaterial surfaces is a physico-chemical property of the major area in the realization / optimization of cell adhesion and proliferation. Thus, the surface properties of the materials are essentially to be analyzed to determine interactions with host tissues [1-7].

The value of the contact angle of the water in contact with the TMZT alloys was determined by the static method of the drop of liquid on a flat surface (Sessile Drop Method) [22].

In order to measure the contact angle of the TMZT alloys, samples of 10 mm x 10 mm x 5 mm were used.

The degree of wetting of the alloys was determined by the contact angle (θ) method, thus determining the hydrophilic or hydrophobic character of the investigated alloys.

Research in the field says that materials with moderately hydrophilic surfaces improve adhesion, cell growth and ultimately, biocompatibility, and a hydrophobic surface can lead to decreased adhesion and loss of biocompatibility.

The Kyowa DM-CE 1 was used to measure the contact angle by placing a drop of water using a microsurge with the 4 microlitre droplet volume on the sample surface.

Ten measurements of the contact angle (θ) were made for each experimental alloy, and the value presented is the average of the measured measurements with a maximum error of $\pm$ 1°. The average contact angle for each alloy is shown in Table V.12 and is represented graphically in Figure V.55 [4].

Table V.12. Values of contact angle with water at the surface of TMZT alloys.

Alloy	Ti15Mo 7Zr5Ta	Ti15Mo 7Zr10Ta	Ti15Mo 7Zr15Ta	Ti20Mo 7Zr5Ta	Ti20Mo 7Zr10Ta	Ti20Mo 7Zr15Ta
Liquid used	water					
Contact angle (degrees)	73.61	45.64	57.93	56.01	70.72	56.11

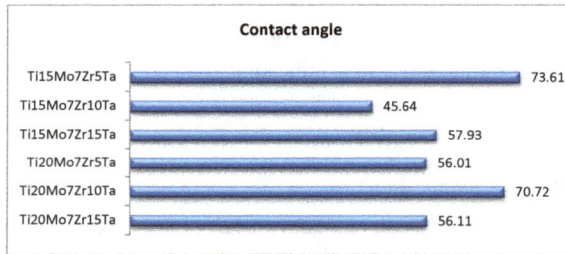

Figure V.55. Graphical representation of contact angle values of experimental TMZT alloys with water.

The water drop on the sample surface was recorded using a digital camera.

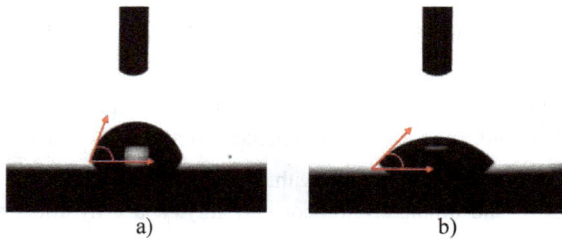

Figure V.56. Images of water droplet on the surface of TMZT alloys: a) Ti15Mo7Zr5Ta, b) Ti15Mo7Zr10Ta.

The images obtained were analyzed later through the Famas program. Figure V.56-58 shows the water droplet image on the surface of the TMZT alloys. These differ depending on the contact angle the water forms on the surface of the sample.

Figure V.57. Images of water droplet on the surface of TMZT alloys:
a) Ti15Mo7Zr15Ta, b) Ti20Mo7Zr5Ta.

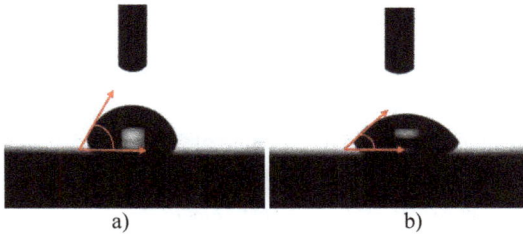

Figure V.58. Images of water droplet on the surface of TMZT alloys:
a) Ti20Mo7Zr10Ta, b) Ti20Mo7Zr15Ta.

All investigated alloys have a contact angle of less than 90°, thus exhibiting a hydrophilic character, which means high cell adhesion to the surface of the alloys.

From the data obtained for the analyzed TMZT surfaces it results that the highest arithmetic mean of the alloys is recorded at the water contact angle on the Ti15Mo7Zr5Ta alloy surface and the lowest at the Ti15Mo7Zr10Ta alloy, this alloy having a more pronounced hydrophilic character.

From the results obtained, it was observed that alloys with a percentage of Mo = 15% present the best values, except the Ti15Mo7Zr5Ta alloy.

The values of TMZT alloys obtained as a result of surface analysis by moistening are certified that titanium alloys have a good interaction with living tissues. Thus, we have the certainty of good cellular adhesion, which could allow the "in vitro" analysis to determine good biocompatibility for the use of alloys developed in the human body [1-7].

V.4.2 Evaluation of Cytotoxicity

A method successfully used for assessing the biocompatibility of metal alloys is that based on the effects of cell cultures, this method becoming the primary way of preliminary testing of in vivo biocompatibility studies for the analyzed biomaterials.

Cell viability for TMZT alloys was tested by the MTT (Tetrazolium Salt Method) [23, 24] test using a tetrazolium salt (3- [4.5-dimethylthiazol-2-yl] 2.5 diphenyltetrazolium bromide) a yellow color and is capable of penetrating the cells.

To evaluate cytotoxicity, 3 metal samples of TMZT alloys, cut to size 5 mm x 5 mm x 3 mm, then prepared specifically for the MTT assay and placed into wells with wells together with the culture medium. Cell viability was expressed as a percentage by reference to control wells (wells containing complete culture medium).

Samples of specified sizes were prepared by exposing them for 30 minutes on each side to the UV action in a transiluminator. Subsequently, they were put in 24-well culture plates for cell sowing for the cell viability study. The prepared samples were kept under sterile conditions in an incubator at a temperature of $37°C$, in a 5% CO_2 atmosphere and a high relative humidity (> 95%).

Preparation of cell cultures consisted in selection of HOS-human osteosarcoma cells (CLS, Eppelheim, Germany) having an osteoblastic phenotype. The cells were thawed and suspended in 20 ml of MEM culture medium (Dulbecco's modified Eagle's medium) supplemented with 10% FBS, 2% L-glutamine and 1% antibiotic (penicillin-streptomycin) in flasks culture (parallelepipedic plastic container with a surface area of 75 cm2). Affinity was reached at 72 hours, after which the culture medium was removed, the adherent cells were washed with PBS (Phosphate Buffered Saline Solution), then cleaved with an EDTA trypsin solution centrifuged and resuspended in 1 ml complete culture medium and subsequently counted with a Neubauer counting chamber. From the 1 ml (medium + cell) work slurry, 4 x 75 cm^2 flasks were seeded and kept in a humidified incubator at $37°C$ and 5% CO_2. After 3 days the confluence in these flasks was complete; the cells were trypsinized, washed with PBS, centrifuged, resuspended in 1 ml of complete medium and counted for put on 24-well cell culture plates.

The study of the viability of HOS cells co-incubated with samples belonging to TMZT alloys consisted of assessing cell viability at 3 and 9 days respectively by seeding the cells after passage on the metal samples in 24-well plates of 1x105cells / well / 1ml MEM complete (MEM with 10% FBS and 1% antibiotic-antimycotic), incubated in humid atmosphere at 37°C and 5% CO_2. For conclusive results, cells were seeded in wells containing only complete culture medium (referred to as Control wells), thus comparing with the results of samples from TMZT alloys. For the 3-day test, after 24 hours the medium was removed by aspiration; washing with PBS and adding completely fresh medium was performed. For the 9-day test, the culture medium supplemented with the extract was first changed after 24 hours and then at 48-hour intervals. For this, the medium was removed by aspiration; washing with PBS and adding freshly complete medium (Figure V.59) was performed [4].

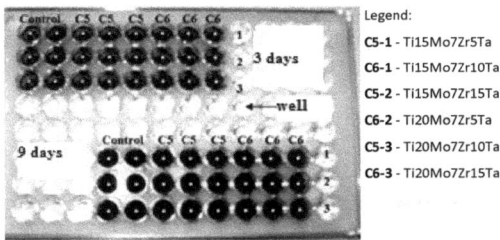

Legend:
C5-1 - Ti15Mo7Zr5Ta
C6-1 - Ti15Mo7Zr10Ta
C5-2 - Ti15Mo7Zr15Ta
C6-2 - Ti20Mo7Zr5Ta
C5-3 - Ti20Mo7Zr10Ta
C6-3 - Ti20Mo7Zr15Ta

Figure V.59. Result of MTT cell viability test steps [1].

Values from the HOS cell viability study coincubated with metallic samples from TMZT alloys are presented in Table V.13.

Table V.13.Results of the cell viability test of HOS cells coincubated with TMZT metal alloy samples.

Alloy	Viability – 3 days (%)	Viability – 9 days (%)
Control	100.00	100.00
Ti15Mo7Zr5Ta (C5-1)	72.55	66.55
Ti15Mo7Zr10Ta (C6-1)	78.09	75.32
Ti15Mo7Zr15Ta (C5-2)	67.71	61.02
Ti20Mo7Zr5Ta (C6-2)	67.56	72.18
Ti20Mo7Zr10Ta (C5-3)	78.43	65.65
Ti20Mo7Zr15Ta (C6-3)	87.86	75.40

The values recorded for TMZT alloys for the 3 and 9-day intervals presented in Table VII.2 are expressed as a percentage of the control wells. Cell viability recorded for 3 days of coincubation of HOS cells with TMZT alloys ranged from 67.56% (Ti20Mo7Zr5Ta) to 87.86% (Ti20Mo10 Zr 15Ta) as compared to control wells. Cell viability recorded for 9 days of co-incubation of TMZT alloys ranged from 61.02% (Ti15Mo7Zr15Ta) to 75.40% (Ti2oMo7Zr15Ta) relative to control wells.

Figure V.60 a and Figure V.60b graphically show the cell viability result for TMZT alloys at 3 and 9 coincubation days, respectively.

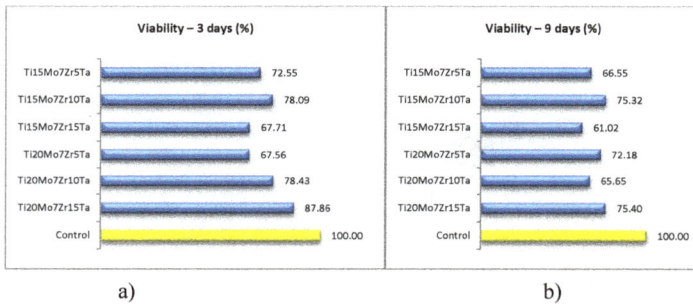

a)   b)

Figure V.60. Results of the MTT test for the cellular viability study of HOS cells coincubated with TMZT alloys: a) 3 days of coincubation; b) 9 days of coincubation.

In Figure V.61 is represents cell viability graphically compared after 3 days and 9 days of coincubation for TMZT alloys.

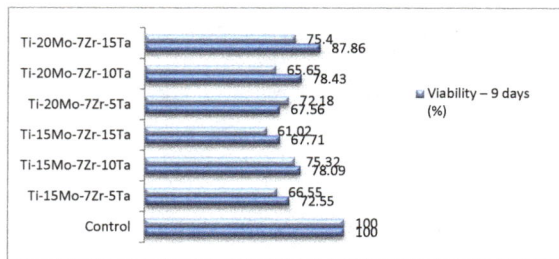

Figure V.61. The results of the MTT test for the cellular viability study of HOS cells co-incubated with the TMZT alloys. The results show comparative the resulting cell viability after 3 days and 9 days coincubation for the TMZT alloys

The obtained results reveal a slightly lower cell viability after 3 days for Ti15Mo7Zr15Ta and Ti20Mo7Zr5Ta (Figure V.60a) and slightly higher for the Ti20Mo7Zr15Ta alloy compared to the other alloys studied.

The results obtained after 9 days recorded in the TMZT alloys a slightly lower viability level for Ti15Mo7Zr15Ta and Ti20Mo7Zr5Ta alloys. However, these differences are not very marked and could be attributed to the compositional characteristics of the samples under study (cytotoxic metal ion release or surface energy) or differences in the topographical characteristics of the samples as the samples had different finishing degrees, and randomly, which significantly influences cell anchor and proliferation in vitro.

Based on the above-mentioned data, and in accordance with ISO 10993 (which specifies: non-toxic material-inhibition of cell growth <25% vs. control, low toxicity for inhibition of cell growth between 26-40% versus control) , it can be say about the studied alloys have low toxicity, and the Ti15Mo7Zr10Ta and Ti20Mo7Zr15Ta alloys are necitotoxic for both studied time intervals [23, 24].

Researches in the field carried out on pure metals (Ag, Al, Cr, Cu, Mn, Mo, Nb, V, Zr) by cell viability tests and compared to pure titanium made a classification of metal toxicity as follows: Ti> Mo> Nb > Zr> Cr> Mn> V> Ag> Al> Cu. Numerous studies indicate that titanium alloys release lower amounts of metal ions in the body than those containing aluminum and vanadium [10, 12].

In the case of TMZT alloys analyzed, their low cytotoxicity was due to the presence of alloying elements (Mo, Zr and Ta), which contribute to the formation of stable oxides.

Based on the results, we can say that TMZT alloys have adequate cell viability and can be considered potential candidates for medical applications.

The values determined for the angle of contact of the TMZT alloys are hydrophilic, which signifies a high adhesion of the cells to the surface of the alloys, is Ti15Mo7Zr10Ta alloy, the alloy of a pronounced hydrophilic character or the less pronounced hydrophilic alloy, is Ti15Mo7Zr5Ta .

Cellular viability assessment studies of HOS cells co-cultivated with metallic samples from TMZT alloys have highlighted that Ti15Mo7Zr10Ta and Ti20Mo7Zr15T alloys are non-toxic for both of the time intervals studied, the other alloys falling within a low cytotoxicity range.

However, the level of viability cannot be fully attributed to the toxic nature of the alloys, and complementary studies complementing aspects of alloy-cell interaction such as zeta potential, microgeometric and topographical topography (results from cutting or corrosion) are necessary.

References

[1]   M.S. Cercel (cas. Bălțatu), Contribuții privind îmbunătățirea proprietăților aliajelor de Ti-Mo destinate aplicațiilor medicale – teză de doctorat, Iași, 2017.

[2]   M.S. Bălțatu, P. Vizureanu, M. Benchea, M.G. Minciună, A.C. Achiței, B. Istrate, Ti-Mo-Zr-Ta Alloy for Biomedical Applications: Microstructures and Mechanical Properties, Key Engineering Materials, 750 (2018)184-188. https://doi.org/10.4028/www.scientific.net/KEM.750.184

[3]   M.S. Bălțatu, P. Vizureanu, V. Geantă, C. Nejneru, C.A. Țugui, S.C. Focșăneanu, Obtaining and Mechanical Properties of Ti-Mo-Zr-Ta Alloys, IOP Conference Series: Materials Science and Engineering, 209 (2017) 012019. https://doi.org/10.1088/1757-899X/209/1/012019

[4]   I. Bălțatu, P. Vizureanu, F. Ciolacu, D.C. Achiței, M.S. Bălțatu, D. Vlad, In Vitro study for new Ti-Mo-Zr-Ta alloys for medical use, IOP Conf. Ser.: Mater. Sci. Eng., 572 (2019) 012030. https://doi.org/10.1088/1757-899X/572/1/012030

[5]   M.S. Bălțatu, P. Vizureanu, V. Goanță, C.A. Tugui, I. Voiculescu, Mechanical tests for Ti-based alloys as new medical materials, IOP Conf. Ser.: Mater. Sci. Eng., 572 (2019) 012029. https://doi.org/10.1088/1757-899X/572/1/012029

[6]   M.S. Baltatu, P. Vizureanu, T. Balan, M. Lohan, C.A. Tugui, Preliminary Tests for Ti-Mo-Zr-Ta Alloys as Potential Biomaterials, Book Series: IOP Conference Series-Materials Science and Engineering, 374 (2018), 012023. https://doi.org/10.1088/1757-899X/374/1/012023

[7]   A.V. Sandu, M.S. Baltatu, Nabialek M., Savin A., Vizureanu P., Characterization and Mechanical Proprieties of New TiMo Alloys Used for Medical Applications, *Materials* 2019, *12*(18), 2973. https://doi.org/10.3390/ma12182973

[8]   M. Bibu, Metode și tehnici de analiză structurală a materialelor metalice, Ed. Universității „Lucian Blaga", Sibiu, 2000.

[9]   C. Popa, V. Cândea, V. Șimon, D. Lucaciu, O. Rotaru, Știința biomaterialelor, Ed. U.T. Press, Cluj-Napoca, 2008.

[10]   Q. Chen, G.A. Thouas, Metallic implant biomaterials, Materials Science and Engineering R, 87 (2015) 1–57. https://doi.org/10.1016/j.mser.2014.10.001

[11]   L.I. Linkow, Prefabicated mandibular prostheses for intraosseous implants, J Prosthet Dent., 20 (4), 1968, p. 367–375. https://doi.org/10.1016/0022-3913(68)90234-5

[12] M. Geetha, A.K. Singh, R. Asokamani, A.K. Gogia, Ti based biomaterials, the ultimate choice for orthopaedic implants - A review, Mater. Sci., 54 (2009) 397-425. https://doi.org/10.1016/j.pmatsci.2008.06.004

[13] R. Chelariu, G. Bujoreanu, C. Roman, Materiale metalice biocompatibile cu baza titan, Ed. Politehnium, Iaşi, 2006.

[14] M.G. Minciună, Contribuţii privind îmbunătăţirea proprietăţilor aliajelor de cobalt utilizate în aplicaţii medicale - Teză de doctorat, Iaşi, 2014.

[15] A.C. Bărbînţă, Îmbunătăţirea proprietăţilor aliajelor de Ti-Nb-Zr-Ta utilizate la fabricarea protezelor ortopedice - Teză de doctorat, Iaşi, 2003.

[16] R.A. Roşu, Metode de obţinere şi de prelucrare a biomaterialelor pentru proteze umane - Teză de doctorat, Universitatea "Politehnica", Timişoara, 2008.

[17] J.R. Davis, Handbook of materials for medical devices, ASM International, United States of America, 2003.

[18] G. Lütjering, J.C. Williams, Titanium-Second Edition, Springer Science + Business Media, Germany, 2000.

[19] G. Zecheru, G. Drăghici, Elemente de ştiinţe şi ingineria materialelor, vol. 1 şi 2, Ed. ILEX şi Ed. Universităţii din Ploieşti, 2001.

[20] I. Voiculescu, I.M. Vasile, E.M. Stanciu, A. Pascu, Ştiinţa şi ingineria materialelor, Indrumar de laborator, Ed. Lux Libris, Braşov, 2015.

[21] M.S.Bălţatu, P. Vizureanu, R. Cimpoeşu, M.M.A.B. Abdullah, A.V. Sandu, The Corrosion Behavior of TiMoZrTa Alloys Used for Medical Applications, Revista de Chimie, 67(10), 2016, pp. 2100-2002.

[22] D.Y. Kwork, C.N.C. Lam, A. Li, A. Leung, R. Wu, E. Mok, A.W. Neumann, Measuring and interpreting contact angles: a complex issue. Colloids and surfaces A., 142 (1998) 219-235. https://doi.org/10.1016/S0927-7757(98)00354-9

[23] T. Mosmann, Rapid colorimetric assay for cellular growth and survival: application to proliferation and cytotoxity assays, J Immunol Methods, 16 (65), 1983, pp. 55-63. https://doi.org/10.1016/0022-1759(83)90303-4

[24] Use of International Standard ISO 10993-1, "Biological evaluation of medical devices - Part 1: Evaluation and testing within a risk management process".

CHAPTER VI: Optimization of Future Ti-based Alloys Biomaterials

A biomaterial may have all the mechanical, physical and chemical characteristics required by a medical application, but upon contact with biological media, including the human body, it finds particular physiological conditions with which it interacts through specific processes such as ion and fluid diffusion, drainage lymphatic, blood circulation, but also through less physiological mechanisms (local and systematic) less predictable. These specific reactions make the material to be tolerated by the environment. Moreover, in the same human body, these mechanical, physicochemical and physiological conditions vary in a fairly wide field.

Regardless of the medical application, a biocompatible material must meet the following requirements:

- to not be toxic and to not contain filter products;
- to not cause allergic, carcinogenic, teratogenic effects (generated by morphological abnormalities);
- to not cause rejection by the body;
- not to alter the composition of the blood and not to disrupt the mechanism of coagulation;
- to not change the biological pH;
- to not cause sedimentation in tissues and biodegradations;
- do not contain hydrophilic or hydrophobic sites that promote cellular adhesion and adhesion;
- if new alloys are not subject to biological acceptance criteria by the animal body, they cannot be placed in the living organism, no matter how appropriate the properties of biomaterials are [1-10].

If new alloys do not fulfill the biological acceptance criteria by the animal body, they cannot be placed in the living organism, no matter how appropriate the properties of biomaterials are.

The biocompatibility of Ti-based alloys is influenced by alloying elements. Table VI.1 lists the main alloying elements of titanium and its influence.

Titanium-Based Alloys for Biomedical Applications   Materials Research Forum LLC
Materials Research Foundations **74** (2020)   https://doi.org/10.21741/9781644900796

Table VI.1. Critical analysis of the biological aspects of the main alloying elements for titanium-based biomedical alloys [1].

Chemical element	Biological aspects
Molybdenum	- the human body contains about 0.07 mg of molybdenum, higher concentrations occur in the liver and kidneys and lower concentrations in the vertebrae; - is an essential element for a number of important enzymes of cellular metabolism; - is less toxic than many other metals (Co, Cr and Ni).
Tantalum	- has no known biological role and is non-toxic; - has excellent corrosion resistance in a large number of acids; the corrosion resistance of tantalum is approximately the same as that of glass; - is among the most biocompatible metals used for implantable devices.
Zirconium	- is found in the body on average 1 mg; - does not play any biological biological role in humans; - the daily intake of zirconium is about 50 mg; - persistent exposure to zirconium tetrachloride leads to increased mortality in rats and guinea pigs, and in decreasing hemoglobin and red blood cells in dogs; - zirconium metal has the highest biocompatibility in the body of all metals and zirconium compounds have low toxicity.
Silicon	-an element found in the natural bone; - in addition to their positive role in the development and proliferation of bone cells, it also provides a porous morphology and structure to develop bone cells (osteoblasts).
Aluminium	- presents acute toxicity at very high doses; - is associated with changes in blood-brain barrier function and brain neurotoxicity; - as with other metals, the toxicity of aluminum is also a major problem in people with kidney disease; - there are studies showing that excessive exposure to aluminum may increase the risk of breast cancer and other neurological conditions such as Alzheimer's disease.
Niobium	- a recent study shows that researchers have found that niobium ions are toxic, along with cobalt, capable of inducing DNA damage and can cause immune cell death; - this element should be treated with care, especially when used with multiple alloying elements.
Vanadium	- Big vanadium doses are toxic to humans, but scientists believe we may need this element in very small quantities for normal bone growth.

Experience in obtaining and characterization of new materials for biomedical applications lead to very good results for the team research: (more than 20 papers and several chapters in books published) with experience in invention elaboration (over 30 patents - awarded at international exhibitions) is the premise of obtaining very good results for the team research.

Our team is focused to implementation and the obtaining titanium based biomaterials with improved characteristics for health applications, as well as the development of technology for its production, from one hand. From other hand, the broadened knowledge on the biomaterial corrosion behaviour and on technological issue for medical alloy production will benefit research society in the efforts to develop biomaterials which fully mimic the behaviour of living tissues.

The expected impact of the recipes results could be considered from different points of view:

- social impact–increased durability of prosthesis based on the developed biomaterial, its biocompatibility, low allergic risk, will result in faster recovery, lower medical treatment cost and improved well-being and health status of the patients.

- impact on the environment– the biomaterial will by obtained by unconventional technologies, without noxious solutions or polluting materials and techniques, but including non-aggressive and energy-saving procedures.

- scientific impact, through the study on a novel combination of elements of Ti-based alloys, the broadened knowledge on the biomaterial corrosion behaviour and on technological issue for medical alloy production will benefit research society in the efforts to develop biomaterials which fully mimic the behaviour of living tissues. The developed biomaterials and technology for its production will be promoted with the participation at international exhibitions, fairs and research conferences, patenting, etc.

- economic impact – by using unconventional (argon protection) technology for metallurgical synthesis, which causes (in general) a decrease in energy consumption, increased economic efficiency will be expected.

The research and development of new-generation metallic biomaterials and their novel fabrication processes will continue. The new generation of biomaterials will overcome the limitations of the cobalt and stainless steel alloys as: high elasticity model and low corrosion resistance, as well as the biocompatibility which depends on the immunologic specificity of each patient. The research aims at functional and economical improvement of biomedical devices to expand their applications to soft-tissue replacement and tissue engineering, in addition to hard-tissue replacement.

Recently, mechanical biocompatibility of biomaterials has been regarded as an important factor, and the research on the development of β types titanium alloys, has been intensively increased [8-10]. The β type titanium alloys show excellent cold workability and high strength. The strength of β type titanium alloys can be increased with keeping Young's modulus low by cold working after solution treatment even the elongation and reduction area are lowered at low cold work ratio by around 20%. At the same time, a low Young's modulus equivalent to that of cortical bone is required in order to inhibit bone absorption into the implant. The alloying elements which are considered non-toxic and non-allergenic based on the reported data of cell viability for pure metals, polarization resistance and tissue compatibility, are: Nb, Ta, Zr, Sn, Mo, Fe, Hf. Some of these can decrease the rigidity of titanium alloys and stabilize the β phase, and reducing the Young's modulus of the Ti alloy (103-120 GPa) at the value comparable with the cortical bone (10-30 GPa), grain size of as cast titanium alloy decreases significantly with boron addition. The biocompatibility of Ti-based alloys indicates an improvement even over commercial pure Ti, the gold standard, and the strength of titanium-based alloys(with Nb, Ta, Zr, Sn, Mo, Fe) implants is found to be up to 40% higher than the strength of cold-worked, grade IV Ti. Lowering the friction coefficient value (ex. for Ti10%Mo alloy), results in the improved dry wear resistance. The wear particles, wear scar depth and width of the 10%Mo alloy are reported to be lower than that of other Ti-Mo alloys [6, 10].

We have undertaken a research study with the main focus on some new compositions of titanium alloys with improved properties as biomaterials: low density, good corrosion resistance, low elasticity and high biocompatibility and increased tolerance in the human body, as compared to stainless steels or CoCr alloys.

The first system is based upon the characterization and promotion of new titanium alloys Ti-Mo-Nb-Sn containing different concentrations of non-toxic metals (Mo, Nb, Sn). These alloys will synergistically combine the effects of each element involved in the titanium system (Mo = 5-10%; Nb = 5-10%; Sn = 0-2%). The obtained biomaterials based on the studied compositions are expected to show improved functional characteristics as implant materials for health-related application.

Another system is Ti-Mo-Zr-Si (Mo: 15%, Zr: 7%, Si: 0-1%). Silicon is considered biocompatible, being a β-stabilizing element that influences the value of the modulus of elasticity.

Molybdenum is an element with a lower degree of toxicity compared to Co, Ni, Cr and is a β-stabilizing element. Field studies [5-8] have highlighted that titanium alloyed with titanium in varying percentages between 15-20% can reduce the modulus of elasticity leading to adequate mechanical properties. From this context we obtained eight alloys (Ti-Mo-Si), to study influence of molybdenum and silicon on titanium alloys.

Development of a new titanium alloy system like (Ti-Mo-Zr-Ta-Nb) for use in medical applications is also in study. By using the Mo, Ta and Nb stabilizers, the alloys proposed have the advantage of increasing mechanical strength and an elastic modulus close to that of the human bone and simultaneously achieving the best properties to meet these criteria. The Ti-Mo-Zr-Ta-Nb system proposed, improve biomaterials with at least 30% of the biomaterials currently used (C.P. Ti, Ti6Al4V, etc.).

The impact of the expected results could be seen in the obtaining titanium based biomaterials with improved characteristics for health applications, as well as development of technology for its production. On one hand, increased durability of prosthesis, its biocompatibility, low allergic risk, will result in faster recovery, lower medical treatment cost and improved well-being of the patients. While on the other hand, the broadened knowledge on the biomaterial corrosion behaviour and on technological issue for medical alloy production will benefit research society in the efforts to develop biomaterials which fully mimic the behaviour of living tissues.

Compared to the classic alloys used in the medical field – C.P. Ti and Ti-6Al-4V, the new Ti-based biomaterial should have at least 30% improved functional properties as implants which is expected to result in increased resistance and a longer lifetime in human body. The new advanced material will provide better solutions to implants-related issues as faster patient recovery based on non-allergic components, increased implant durability based on improved corrosion resistance in human fluids, patient wellness based on the mechanical properties close to the human tissues. The expected results from such studies offer new possibilities for healthy ageing of the human population.

The advantages of all alloys obtained: they are obtained in a controlled environment, protecting the environment and the health of the population; the possibility of melting the vacuum metal samples under a protective atmosphere by means of a non-consumable mobile electrode of thorium tungsten; uniform alloys created by repeated replication, can be successfully used at the stages of their approval as medical implants, through the possibility of patenting composition and technological transfer to the profile companies; can be reused / recycled as a starting material subsequently, have a low risk of producing carcinogenic effects, allergies, using non-toxic elements.

The proposed alloys are completely NEW in the field, not being reported in any research that addresses complex matrices that includes all these elements simultaneously. The motivation of studying these compositions is to enhance the elasticity and the mechanical resistance without affecting the biological capabilities of the alloy: to increase the plasticity, increasing the mechanical resistance and decreasing the friction coefficient (alloying with Mo), increasing the fatigue resistance and ductility (alloying with Nb), reducing Young modulus (alloying with Sn, Si) and refining grains (alloying with Zr). The planned alloying elements induce, usually, a relatively low elastic modulus and excellent resistance to corrosion due to the oxide surface layers that allows implant fixation and ensure compatibility [1-7].

The newly developed alloys will be tested through chemical, structural, thermal, surface and mechanical analysis. The following chemical and physicochemical analysis are envisaged: Chemical composition (SEM/EDX or ICP-OES – to obtain the data on the chemical composition of the obtained alloys; Structural characterization (Optical microscopy, X-ray diffraction XRD, TEM) – to obtain data on microstructure, phase composition, crystallographic orientation, texture, etc.; Mechanical characterization (Hardness Tests, Indenting Tests) - highlights the mechanical properties of the elaborated alloys: hardness, elasticity etc.; Corrosion resistance: Linear and cyclic polarization - determines the stability of the proposed alloys in the human fluid; Surface characterization (Contact angle (θ)) - involves measuring the contact angle of the alloy surface to achieve / optimize cell adhesion and proliferation; Biocompatibility tests: is required to assess the in vitro events occurring during cells-materials co-incubation and/or interactions.

References

[1]   Q. Chen Q, G.A. Thouas, Metallic implant biomaterials, Materials Science and Engineering R, 87 (2015) 1–57. https://doi.org/10.1016/j.mser.2014.10.001

[2]   R. Chelariu, G. Bujoreanu, C. Roman, Materiale metalice biocompatibile cu baza titan, Ed. Politehnium, Iaşi, 2006.

[3]   I. Antoniac, Biomateriale metalice utilizate la executia componentelor endoprotezelor totale de sold, Ed. Printech, Bucureşti, 2007.

[4]   D.M. Bombac, M. Brojan, P. Fajfar, F. Kosel, R. Turk, Review of materials in medical applications, Materials and Geoenvironment, vol.54(4), 2007, pp. 471-499.

[5]   C.N. Elias, J.H.C. Lima, R. Valiev, M.A. Meyers, Biomedical applications of titanium and its alloys, Biological Materials Science, 2008, pp. 46- 49. https://doi.org/10.1007/s11837-008-0031-1

[6]   M. Niinomi, Mechanical properties of biomedical titanium alloys, Mater. Sci. Eng., A, 243 (1998) 231-236. https://doi.org/10.1016/S0921-5093(97)00806-X

[7]   L.I. Linkow, Prefabicated mandibular prostheses for intraosseous implants, J Prosthet Dent., 20(4), 1968, pp. 367–375. https://doi.org/10.1016/0022-

3913(68)90234-5

[8] L.I. Linkow, The blade vent-a new dimension in endosseous implantology, Dent Concepts, 11 (2), 1968, pp.3-12.

[9] M. Geetha, A.K. Singh, R. Asokamani, A.K. Gogia, Ti based biomaterials, the ultimate choice for orthopaedic implants - A review, Mater. Sci., 54 (2009) 397-425. https://doi.org/10.1016/0022-3913(68)90234-5

[10] M.S. Cercel (cas. Bălţatu), Contribuţii privind îmbunătăţirea proprietăţilor aliajelor de Ti-Mo destinate aplicaţiilor medicale – teză de doctorat, Iaşi, 2017.

CHAPTER VII: Destination of Ti-based Alloys used in the Human Body

A biomaterial can be defined as any material used to make devices intended to replace a part or function of the body in a safe, efficient, economical and physiologically acceptable manner. Researchers [1-7] considered at the outset that wood and bone are included in the biomaterials category, the term being incorrectly used because they are in fact biological materials. One of the definitions of biomaterials can be expressed as follows: a synthetic material used to replace a part of a living system or contacting a living tissue. The Council of the Clemson University for Biomaterials in South Carolina (Columbia) formally defined a biomaterial as "a systemic and pharmacologically inert substance designed for implantation, or incorporated into a living organism." A biological material, unlike biomaterials, is a material such as epidermis or bone tissue arteries produced by a biological system.

At present, a wide variety of materials and devices are used to treat traumas or conditions. Classical examples include materials for screws, obturators, catheters and plates for bone reconstruction. In the case of medical applications, the main purpose of using biomaterials is to improve health by restoring functions of living tissues, and it is essential to study and understand the relationships between the properties, functions and structures of biological materials.

The term biometal does not have its own definition, however, by this term is meant a metallic material used as a biomaterial. Biometrics are used in the form of medical devices, prostheses, etc. These are defined as follows [1-8]:

- medical device - any instrument, equipment, appliance, software, material or other article used separately or in combination, including software specifically intended by the manufacturer for diagnosis and / or therapeutic purposes and necessary for the proper operation of the device medical, aiming at:

a) diagnosis, prevention, monitoring, treatment or amelioration of a condition;

b) diagnosis, monitoring, treatment, amelioration or compensation of a lesion or disability;

c) control of the concept and which does not fulfill its principal purpose for which it has been intended in the human body or by its pharmacological, immunological or metabolic means but whose operation can be assisted by such means;

d) investigation, replacement or modification of anatomy or physiological process;

- Protesis - is a device designed to physically replace a member, organ or body tissue;

- Implant - is a medical device that can be made of one or more biomaterials and is placed in the human body, partially or totally, for a significant period of time;

- Artificial organ - is a medical device that can partially or totally replace the function of a body organ;

- Surgical alloy - is a metallic material used as a biomaterial;

- Biocomposite - is a material that results from the combination of two or more biomaterials.

Successful use of advanced metallic material is largely dependent on four major factors: properties, biocompatibility, patient's health, and the competence of the surgeon using the alloy.

The characteristics of a biocompatible material must be in full compliance with its intended purpose. Thus, the use of biocompatible materials implies:

- Keeping the life or viability of a certain human organ (heart valves, cardiovascular filters, cerebral sutures);

- Removal of diseased or destroyed parts of the human body, which have lost their functionality due to diseases or traumas (endoprostheses, dental implants);

- Assisting in the healing of parts of the human body (sutures, bone plaques, bioresorbable);

- Regulation of functional abnormalities (cardiac pacemaker);

- Improvement of human functions (intraocular lenses);

- Correction of cosmetic problems (artificial skin, breast implant);

- Diagnosis and treatment aid (catheters, drainage tube).

Also, depending on how the biocompatible material is implanted in the body, the following can be encountered [8]:

- Implanted in the body (endoprostheses, centromedullary rods, plaques);

- Inside a cavity of the human body (implants);

- Outside the human body, with the possibility of accessing internal tissues (external fixators for osteosynthesis).

One of the main conditions for classifying biomaterials is that they have to have a high degree of chemical inertia. With the development of biomaterials, problems have arisen regarding the type of material used for implants, leading to premature loss of implant functions due to insufficient mechanical damage, corrosion or biocompatibility. Biocompatibility and biofunctionality are decisive factors for the use of metallic materials in medical devices and implants [9-10].

Other features that may be important in the action and structure of a biomaterial implant are adequate mechanical properties, such as breaking-compressive strength, hardness and resilience [11]. Metal implants for hip interventions are designed to hold the hip joints that have degenerated and become painful or that (rarely found) have been deformed since birth or damaged during an accident. A part of the joint is called the femoral head and the other part of the joint is called the acetabular cup, an example of the implant is represented in Figure VII.1.

The implants produced by the main companies are titanium based. This is a very light metal and extremely well tolerated by the human body. Titan has been using for many years in other branches of medicine as well: repair surgery, orthopedics, etc.

The first report on the use of titanium in the human body dates back to 1940. In an experimental study on cats, bone proliferation was directly observed on the metal implant. There is no clue on the number of cases and treatment, so it is difficult to tell who was the first to promote titanium implants in humans. Titan has attracted the attention of the medical world through its particularly advantageous properties: biocompatibility, low thermal conductivity, low density, corrosion resistance, odorless and insipid character, the cost price of the material being four times lower than gold [12-13].

The introduction of this metal into medical applications has raised many technological problems due to the special chemical reactivity and difficult processing conditions. For these reasons, the first ones that have entered the therapeutic arsenal of medical specialties were those parts obtained industrially, through rigorously controlled technological processes and generally cold processed (implants, pre-fabricated corono-radicular devices, endodontic cones, wire and orthodontic devices, osteosynthesis plates for oro-maxillofacial surgery, instrumental, etc.) [14].

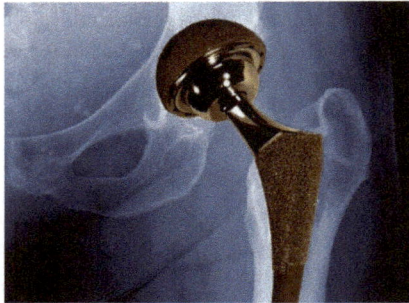

Figure VII.1. The image of an implant [11].

The advantages of using implants are because they have been using for a very long time, in the last 10 years this field has seen incredible growth. There are currently over 1000 different systems around the world.

1. The old method of treatment consists of bridges or prostheses. Both of these methods affect the integrity of teeth that are still present. The main advantage of implant treatment is that it leaves the neat neighbors teeth. They will not be affected in any way.

2. Another advantage is given by the comfort achieved with implant treatment is superior to that offered by different prostheses.

3. Even if there are no longer teeth and a prosthesis is already in use for many years, you can benefit from implant treatment and these will stabilize the prosthesis very well and in some situations it is even possible to give up the prosthesis completely.

4. Using the implants requires you to maintain a particularly good hygiene.

In all cases, implant treatment cannot be used, and in order to achieve this, a set of radiographs and a set of analyses are needed.

Metallic materials used in the medical field are produced and used in a direct relationship and closely related to their physical, chemical and biological properties, valued according to world standards. The properties of a material used in medicine are not only appreciated by a single character but by the behavioral characteristics in the biological environment and by assessing all the aspects of these materials subjected to different demands and different attempts which give the general characteristics of these materials. Actual, science and technology have grown considerably, also making considerable progress in

Materials Research Forum LLC
https://doi.org/10.21741/9781644900796

the field of biomaterial research; they have diversified and the spheres of use have widened, including infrastructure and infra-biology.

The use of biomaterials in medicine is strongly dependent on the rational dimension of mechanical strength, physical, chemical and biological characteristics. Today, testing laboratories have, for the quality control of a biomaterial, tests and qualified personnel, each unit of metallic medicine producing units has a testing service in line with international standards. The need to standardize the qualities and tests of different materials for medical use has required the establishment of rules in the medical field, norms developed and established by different competent form, such as:

✓ International Organization of Standards (I.S.O.);

✓ German Institute of Norms (D.I.N);

✓ American Medical Association (A.M.A.) etc.

Standardization of medical products has its origins in the beneficiaries' protection priorities since 1928, and since 1950 British and even Australian criteria have emerged, today more than 60 countries have collaborated in the development and application of I.S.O standards. It should be noted that I.S.O. are not binding on all countries, but are binding by law for those countries that have joined these associations, and which ensures the fulfillment of the most severe requirements, and the World Health Organization (OMS) is also involved in their elaboration.

The application of the norms for metallic medical products should be done in a standardized, reproducible manner, in view of achieving the following objectives:

- classification of medical materials according to the nature of the material;
- establishing the methods of physical, chemical and biological testing of medical materials;
- establishing the quality of the medical materials, ensuring consumer protection;
- establishing the conditions of delivery, storage and use of each type of material;
- classification of the material in quality groups, providing additional data on the use of each material;
- classification of the material by applying a class indicator that facilitates the assessment of quality, to which must be added also marks on the date of manufacture and the duration of the guarantee.

The qualities of a metallic medical material are determined on the basis of tests of unitary charges, which are examined by inspection and physical, chemical and biological testing. At present, there are rules for all metallic materials, which cannot be marketed without the help of one of the institutions empowered in this meaning and above mentioned [18-20].

Titanium alloys must have high biocompatibility, good corrosion resistance and excellent mechanical properties (low density, low Young's modules for use in fields such as orthodontics and orthopedics, but also in cardiovascular and reconstructive purposes. Some titanium alloys receive more attention as biomaterials due to their high specific weight and good corrosion resistance, without allergic problems, and show the best biocompatibility between metallic biocompatible materials (Figure VII.2).

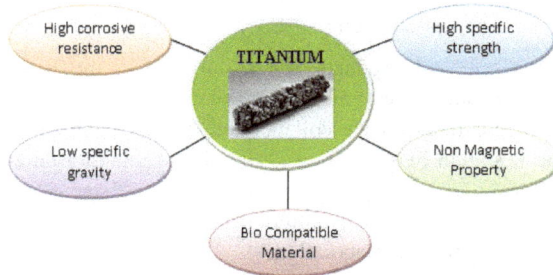

Figure VII.2. Titanium properties for medical use [18-20].

Titanium has been found to be the only metallic biomaterial for bone integration and has a bioactive behavior (greatly improves the quality and duration of use of implants) due to the slow increase of titanium hydrated oxide on the surface of the titanium implant leading to the incorporation of calcium and phosphorus.

The influence of alloying elements in titanium alloys contributes to a wide range of different micro structural and mechanical properties. Thus, the alloying elements are divided into three categories: stabilizers α: C, N2, O2, Al; β stabilizers: V, Nb, Mo, Ta, Fe, Mn, Cr, Co, W, Ni, Cu, Si, H2; neutral elements: Zr, Sn, Hf, Ge, Th [21-36].

According to recent studies, since the 1990s, titanium alloys have been studied and improved with various alloying elements. The authors correlated the values of the mechanical characteristics with X-ray diffraction results and pointed out that low elastic modulus alloys are due to the full presence of the β phase. Types of Ti-based alloys systems in different combination with biocompatible elements: Ti-Al, Ti-Al-Sn, Ti-Al-Zr, Ti-Al-Sn-Cu, Ti-Cu-Zr (α phase), Ti-Al-Mn, Ti-Al-V, Ti-Al-Mo, Ti-Al-Mo-V, Ti-Al-Mo-Cr (α+β phase), Ti-Mo, Ti-Nb, Ti-Ta, Ti-Zr, Ti-Zr-Nb, Ti-Sn-Nb, Ti-Nb-Ta-Zr (β phase), Ti-Mo-Si, Ti-Mo-Zr-Ta, Ti-Al-Zr, Ti-Al-Sn-Cu, Ti-Cu-Zr (α phase), Ti-Al-Mn, Ti-Al-V, Ti-Al-Mo, Ti-Al-Mo-V, [25], Ti-Mo, Ti-Nb, Ti-Ti, Ti-Zr, Ti-Zr-Nb, Ti-Sn-Nb , Ti-Mo-Si (β phase), Ti-Mo-Zr-Ta (β phase) [21-36].

Metallic biomaterials represent approximately 70-80% of all medical implant materials. The main metallic biomaterials are stainless steels, Co-based alloys and Ti-based alloy which show a good biocompatibility.

Metallic biomaterials cannot be replaced with bioceramics at present because are sensitive to stress concentrations which exist around pre-existing defects, such as pores, scratches, or cracks. A biomaterial must have very high biocompatibility, no secondary effects on tissue response. Metallic biomaterials like Ti based alloys, Co based alloys and stainless steel alloys must have a low value density like bone, high mechanical strength, very good fatigue properties and low elastic modulus. The main disadvantage of these metallic biomaterials is the release of metallic ions in the human body which cause harmful effects [19].

Table VII.1 presents mechanical characteristics of the main biomaterials used in medical applications. Titanium alloys shows the best characteristics compared to the cobalt-based alloys and stainless steels, as is presented in Table VII.1.

Table VII. 1. Mechanical characteristics of the main biomaterials [19, 7].

Characteristics	Titanium alloys	Cobalt alloys	Stainless steel
Stiffness	Low	Medium	High
Strength	High	Medium	Medium
Corrosion Resistance	High	Medium	Low
Biocompatibility	High	Medium	Low

Figure I.2 highlights the main applications of biomaterials used in the human body and Figure I.3 shows the main applications of titanium alloys.

Metals are the most used biomaterials for orthopedic implants, and are known for their high resistance to wear, ductility and high hardness [10].

Pure titanium and Ti6Al4V alloy were the first titanium based materials for commercial use, used as biomaterials [5, 16].

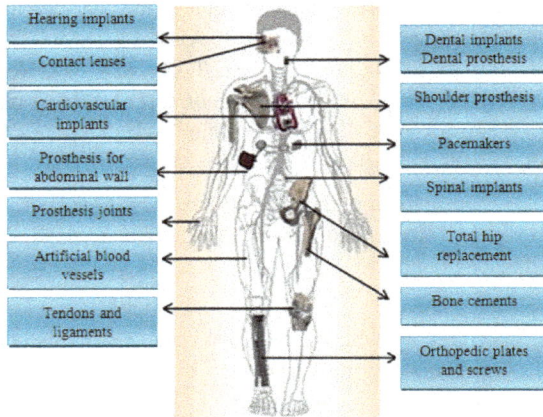

Figure VII.3. Representative example of application of metallic biomaterials.

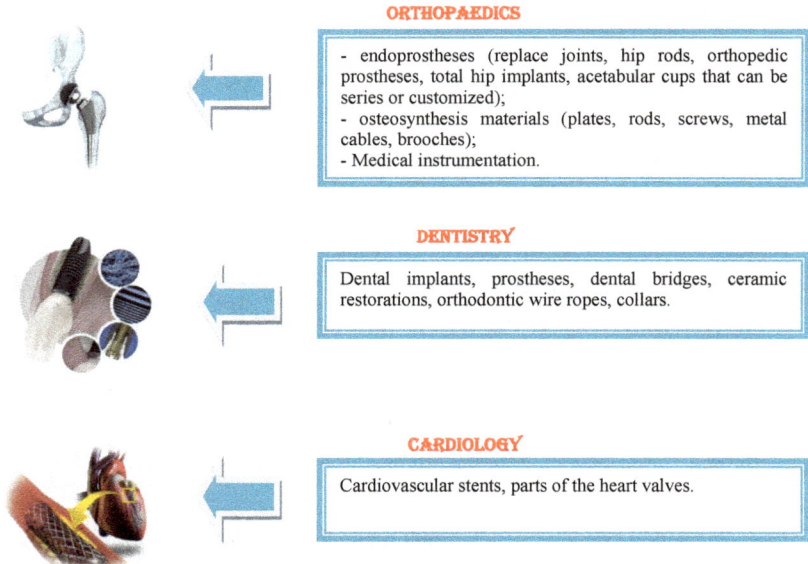

ORTHOPAEDICS

- endoprostheses (replace joints, hip rods, orthopedic prostheses, total hip implants, acetabular cups that can be series or customized);
- osteosynthesis materials (plates, rods, screws, metal cables, brooches);
- Medical instrumentation.

DENTISTRY

Dental implants, prostheses, dental bridges, ceramic restorations, orthodontic wire ropes, collars.

CARDIOLOGY

Cardiovascular stents, parts of the heart valves.

Figure VII.3. Various applications of biomaterials in the human body [19, 20].

Compared to stainless steels and Co-based alloys, titanium alloys have a better biocompatibility [16]. Stainless steels and CoCr alloys are prone to corrosion, releasing metal ions into the body that can cause side effects.

The area in which titanium is widely used is orthopedic by the production of orthopedic hip prostheses (Figure VII.5), shoulder prostheses, orthopedic screws and pads, etc. [19, 20].

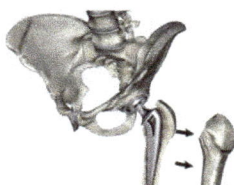

Figure VII.5. Titanium hip orthopaedic prosthesis [38].

In dentistry, titanium was initially used in dental implants (Figure VII.6.), and in recent years applications of titanium alloys have also been diversified in dental prosthetics and orthodontics.

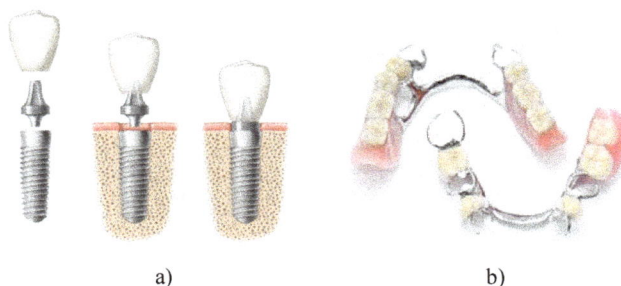

a)   b)

Figure VII.6. Utilisation of titanium in dentistry: a) dental implants, b) skeletal partial prosthesis.

The resistance and hardness of titanium is very close to that of noble alloys.

In cardiovascular surgery, titanium-based materials are commonly used in the form of metallic stents (NiTi, TiNO). The most commonly used material is nitinol, in the form of self-expandable stents, a form memory alloy that retains shape with its predetermined size and configuration (Figure VII.7).

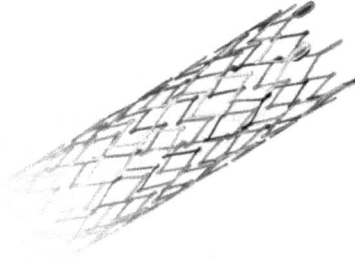

Figure VII.7. Nitinol stent [41]

Ti-based alloy has attractive properties for biomedical applications where the most important factor is biocompatibility therefore its use is indicated in the sphere of the human body. Fields of application of titanium and its alloys in dentistry and implantology are in continuous expansion, and this is highlighted in the table 2, being more used than Co-Cr alloys and stainless steel.

Table VII.2. Medical applications of metallic biomaterials and its frequency [12-14, 19].

	316L stainless steel	Co-Cr-Mo	CP-Ti	Ti-6Al-4V	Ti-13Nb-13Zr	Ti- Mo-Zr-Fe	Ti-15Mo
Bone and joint replacement		X	X	X	X	X	X
Fracture fixation	X	X	X	X	X	X	X
Stents	X						
Surgical instruments	X		X	X			
Dental implants		X	X	X	X	X	
Pacemaker encapsulation		X	X	X	X		
Heart valves		X	X				
Dental restorations		X					

Another aspect for the choice of Ti-based alloys is that Young modulus is closer than of bone tissue (10-30 GPa), beside Co-Cr-Mo and stainless steel. If the Young modulus of the implant is significantly higher than Young modulus of human bone it will be generated the stress-shielding effect, which is the cause for bone failure.

Titanium alloys have to meet certain requirements and take into account some important characteristics. Properties of interest in the use of these alloys are: mechanical properties, biocompatibility and corrosion resistance.

References

[1] M. Dourandish, V. Firouzdor, D. Godlinski, A. Simchi, Materials Science and Engineering A — Structural Materials Properties, Microstructure and Processing, 472 (2008) 338. https://doi.org/10.1016/j.msea.2007.03.043

[2] P.R. Geissler, K.M. McKinlay, A two-part prosthesis for edentulous patients following maxillectomy, Journal of Dentistry, 5(4). 1977, pp. 331-333. https://doi.org/10.1016/0300-5712(77)90126-9

[3] P.R. Geissler, K.M. McKinlay, Aluminium bronze as a denture base material, Pract Dent Rec, 19 (1969) 278-80.

[4] B. Ghiban, Metalic Biomaterials, Editura Printech, 1999.

[5] L.L. Hench, Biomaterials: a forecast for the future, Biomaterials, 19 (1998) 1419–1423. https://doi.org/10.1016/S0142-9612(98)00133-1

[6] L.L. Hench, J.M. Polak, Third-generation biomedical materials, Science, 295 (2002) 1014-1017. https://doi.org/10.1126/science.1067404

[7] A.S. Hoffman, J.E. Lemons, B.D. Ratner, F.J. Schoen, Biomaterials Science: An Introduction to Materials in Medicine. Elsevier Science, 2012.

[8] M.G. Minciună, P. Vizureanu, Materiale metalice avansate pentru aplicații medicale, editura PIM, Iași, 2016, 178pg., ISBN 978-606-13-3529

[9] F. Rupp, L. Scheideler, D. Rehbein, D. Axmann, J. Geis-Gerstorfer, Roughness induced dynamic changes of wettability of aidetched titanium implant modifications. Biomaterials, 25 (2004) 1429-1438. https://doi.org/10.1016/j.biomaterials.2003.08.015

[10] M.H. Kim, S.Y. Lee, M.J. Kim, S.K. Kim, S.J. Heo, J.Y. Koak, Effect of biomimetic deposition on anodized titanium surfaces. J Dent Res, 90, 2011, p. 711-716. https://doi.org/10.1177/0022034511400074

[11] A. Krozer, J. Hall, I. Ericsson, Chemical treatment of machined titanium surfaces. Clin Oral Implant Res, 10 (1999) 204-211. https://doi.org/10.1034/j.1600-0501.1999.100303.x

[12] M. Geetha, A.K. Singh, R. Asokamani, A.K. Gogia, Prog Mater Sci, 54 (2009) 397. https://doi.org/10.1016/j.pmatsci.2008.06.004

[13] G. Lutjering, J.C. Williams, Titanium. 2nd ed. Berlin: Springer; 2007.

[14] V. Milleret, S. Tugulu, F. Schlottig, H. Hall, Alkali treatment of microrough titanium surfaces affects macrophage/monocyte adhesion, platelet activation and architecture of blood clot formation, Eur Cell Mater, 21 (2011) 430-444. https://doi.org/10.22203/eCM.v021a32

[15] A. Peutzfeld, E. Asmussen, Distortion of alloy by sandblasting. Am J Dent., 9 (1996) 65- 66.

[16] C.R. Choi, H.S. Yu, C.H. Kim, J.H. Lee, C.H. Oh, H.W. Kim, et al., Bone cell responses of titanium blasted with bioactive glass particles. J. Biomater Appl, 25 (2010) 99-117. https://doi.org/10.1177/0885328209337345

[17] R. Jeong, C. Marin, R. Granato, M. Suzuki, J.N. Gil, J.M. Granjeiro, et al., Early bone healing around implant surfaces treated with variations in the resorbable blasting media method. A study in rabbits, Med Oral Pathol Oral Cir Bucal, 15 (2010) 119-125. https://doi.org/10.4317/medoral.15.e119

[18] H. Conrad, Prog Mater Sci, 26 (1981) 123. https://doi.org/10.1016/0079-6425(81)90001-3

[19] M.S. Bălțatu, Vizureanu P., Țierean M.H., Minciună M.G., Achiței D.C., Ti-Mo Alloys used in medical applications, Advanced Materials Research, vol. 1128, 2015, p. 105-111. https://doi.org/10.4028/www.scientific.net/AMR.1128.105

[20] M.S. Baltatu, C.A. Tugui, M.C. Perju, M. Benchea, M.C. Spataru, A.V. Sandu, P. Vizureanu, Biocompatible Titanium Alloys used in Medical Applications, Revista de Chimie, 70(4), 2019, 1302-1306. https://doi.org/10.37358/RC.19.4.7114

[21] M.S. Cercel, (Bălțatu), Teza de doctorat, Contribuții privind îmbunătățirea proprietăților aliajelor Ti-Mo destinate aplicațiilor medicale, Iași, 2017.

[22] J. Markhoff, M. Krogull, C. Schulze, C. Rotsch, S. Hunger, R. Bader, Biocompatibility and Inflammatory Potential of Titanium Alloys Cultivated with Human Osteoblasts, Fibroblasts and Macrophages, Materials 52 (10), 2017, pp. 1-17. https://doi.org/10.3390/ma10010052

[23] C.N. Elias, J.H.C. Lima, R. Valiev, M.A. Meyers, , Biomedical Applications of Titanium and its Alloys, Biological Materials Science, JOM (2008) 46-49. https://doi.org/10.1007/s11837-008-0031-1

[24] S. Liliane, et all, "Titanium Alloy Mini-Implants for Orthodontic Anchorage: Immediate Loading and Metal Ion Release," Acta Biomaterialia, 3 (2007) 331–339. https://doi.org/10.1016/j.actbio.2006.10.010

[25] A. Savin, R. Steigmann, N. Iftimie, F. Novy, P. Vizureanu, M.L. Craus, S. Fintova, Nondestructive evaluation of the interface between ceramic coating and stainless steel

by electromagnetic method, 7TH International Conference On Advanced Concepts In Mechanical Engineering, Book Series: IOP Conference Series-Materials Science and Engineering, 147 (2016) 012030. https://doi.org/10.1088/1757-899X/147/1/012030

[26] B. Istrate, C. Munteanu, M.N. Matei, B. Oprisan, D. Chicet, K. Earar, Influence of ZrO2-Y2O3 and ZrO2-CaO coatings on microstructural and mechanical properties on Mg-1,3Ca- 5,5Zr biodegradable alloy, International Conference on Innovative Research, ICIR, 2016, 123393. https://doi.org/10.1088/1757-899X/133/1/012010

[27] D.C. Achitei, M.M.A. Abdullah, A.V. Sandu, P. Vizureanu, On The Fatigue of Shape Memory Alloys, Advanced Materials Engineering And Technology II, Book Series: Key Engineering Materials 594 (2014) 133-139. https://doi.org/10.4028/www.scientific.net/KEM.594-595.133

[28] Information on http://www.supraalloys.com

[29] M. Niinomi, T. Narushima, M. Nakai, Advances in Metallic Biomaterials; Springer: Berlin/Heidelberg, Germany, 2015. https://doi.org/10.1007/978-3-662-46836-4

[30] Y.H. Li, C. Yang, H.D. Zhao, S.G. Qu, X.Q. Li, Y.Y. Li, New developments of Ti-Based alloys for biomedical applications, Materials, 7 (2014) 1709-1800. https://doi.org/10.3390/ma7031709

[31] T.M. Mohsin, A.K. Zahid, N.S. Arshad, Beta titanium alloys: the lowest elastic modulus for biomedical applications: a review, Int. J. Chem. Nucl. Metall. Mater. Eng., 8 (2014) 726-731.

[32] C. Cumpata, N. Ganuta, Theoretic considerations regarding the obtaining of dental implants, Revista română de stomatologie – volumul LVII, 1, 2016.

[33] C. Oldani, A. Dominguez, Titanium as a Biomaterial for Implants, Recent Advanced in Arthroplasty, (2012) 149-162. https://doi.org/10.5772/27413

[34] I. Csaki, G. Popescu, R. Stefanoiu, In situ Reaction Influence on Hybrid Aluminum/Titan Aluminide and Alumina Composite Hardness, Revista de Chimie 64, 7 (2013) 693-696.

[35] I. Ohnuma, Y. Fujita, H. Mitsui, K. Ishikawa, R. Kainuma, K. Ishida, Phase equilibria in the Ti-Al binary system, Acta Materialia 48 (12), 2000, pp. 3113-3123. https://doi.org/10.1016/S1359-6454(00)00118-X

[36] H. Michelle Grandin, S. Berner, M.A. Dard, Review of Titanium Zirconium (TiZr) Alloys for Use in Endosseous Dental Implant, Materials, 5 (2012) 1348-1360. https://doi.org/10.3390/ma5081348

CHAPTER VIII: Exploitation and Trends

VIII.1. Surface Modification

By definition, biocompatible materials do not cause adverse effects to surrounding structures, do not cause inflammation, give rise to allergic reactions, and do not cause cancer. Biomaterials can be defined as "any synthetic substance (metals and alloys, ceramics, polymers, composites, etc.) or natural (collagen, chitosan, cellulose) capable of interacting with biological systems for the treatment, growth or replacement of any tissue, or function of the body ". Biomaterials are elements that have an increased capacity to integrate and be accepted by the biological environment in which they are implanted, being able to come into contact with natural fluids and human tissues without causing adverse reactions and with very few undesirable effects.

Titanium and its alloys have been extensively studied lately due to the special properties of these special metallic materials. The surface treatment of titanium or its alloys is closely related to their properties and the knowledge of the main characteristics is absolutely necessary for their further processing.

Biocompatibility of titanium is a consequence of the presence of superficial oxide layer. Titanium forms a series of stable oxides such as TiO_2, TiO and Ti_2O_3, the most widespread being TiO_2.

From this point of view, bicompatibility depends fundamentally on:

1 - Adhesion of biomolecules: there are weak links at the surface of the implant, with a continuous exchange of biomolecules and strong, even irreversible bindings, which implies the establishment of a dynamic balance of the bonding forces at the interface.

2 - Chemical Process: it has been found that the thickness of the oxide layer increases more rapidly in the implant inserted in the bone than on a similar surface exposed to air. Resistance of adhesion between biomolecules and the surface of the implant determines the ratio of molecule replacement to this level. Oxidation layer augmentation occurs under the conditions of hydrogen atoms participation in the formation of hydroxides. Other components that can contribute to the formation or modification of the oxide layer are Ca and P.

3 - The roughness of titanium surfaces, which has two aspects:

- its value is much higher than the atomic dimensions. In this case, it intervenes favorably in the distribution of stresses, and chemical bonds are not influenced by roughness.

- its value is comparable to atomic dimensions. In this case, it influences the quantity and quality of chemical bonds at the interface level, due to the geometric arrangements that can be established.

Surface quality is particularly significant for biomaterials and hence the peculiarity of surface modification techniques for biomaterials. The predominant processes for modifying the surfaces of metallic materials are the dried ones (ion beam, electrons or photons) and the wet ones (which take place in different solutions). Techniques like electro-chemical techniques in wet processes have become increasingly important. Also another technique for improving bone-compatibility such as some proteins or biomolecules (collagen or peptide) is gaining interest. On the other hand, it is also attempted to immobilize biofunctional molecules such as polyethylene glycol (PEG) on the surface of metallic material to control the rate of adsorption of adhesion proteins and cells or bacteria.

VIII.1.2 Dry Surface Modification Processes

Most dry processes use a beam of ions, atoms or photons. Ion-based technologies are useful in engineering domains, particularly in silicon technologies. Fascicle technologies allow the formation of thin films at atomic or molecular levels. The process takes place on the basis of a thermal difference which leads to the possibility of synthesizing non-natural substances (new matter that cannot normally exist).

This beam technology has significantly contributed to the modification of biomaterial surfaces. A classification based on effects on the solid surface involves the formation of a superficial layer (spraying) and material implants (ions, atoms, molecules). When the beam comes into contact with the surface of the material, different attachment, spraying or implantation effects occur depending on the energy of the particles of the beam. By using these effects thin layers or layers of different chemical composition can be formed.

a) Hydroxyapatite coatings (HA)

At present, plasma spraying of HA on metallic materials is used extensively to form HA coils forming the nucleus and the beginning of bone growth and its bindings. In the case of spraying hydroxyapatite on the HA-Ti interface or even the HA layer, it could crack due to the internal stresses that arise due to the low bonding force between the two materials and the strength of the layer actually deposited.

The solubility of ceramic materials increases as their crystallinity decreases. The crystallinity of thin film-formed films is low, so their solubility is high. The crystallinity of HA-coated materials is an important factor because they control the solubility in the human body. Low crystallinity Ti films will dissolve rapidly upon introduction into the human body. One solution is the implantation of Ca ions to increase the bond between the superficial HA film and the Ti substrate, which have a binding role. HA and CaTiO4 coatings have already been performed by spraying with magnetron and the results have been reported in the literature.

The most important property of this ceramic material, hydroxyapatite, is the ability to interact with living bony tissue, forming strong bindings to the bone. It is commonly used for orthopedic and maxillofacial applications, either as a coating for metal implants or as a bone filler [84]. However, the material has some disadvantages - HA is not thermally stable, decomposing at 800-1200°C depending on its stoichiometry. This material has poor mechanical properties (especially low fatigue strength), which means it cannot be used in compact form for applications where the implant is subjected to heavy mechanical stress (e.g. hip joint).

Hydroxyapatite coatings are often used for metallic implants (especially titanium, titanium alloys or stainless steel) in order to modify surface properties. Most applications of HA coatings are for endo-bone and subperiosteal implants and for orthopaedic devices. By coating with a hydroxyapatite layer, the implant benefits both from biocompatibility and has the ability to form chemical bonds with the living bone, as well as it can benefit from the mechanical properties of the biocompatible alloy substrate.

Due to the osteophilic area of hydroxyapatite, the mechanical burden acting on the implant will be transferred to the bone skeleton by helping to combat bone atrophy. Plasma-coated hydroxyapatite-coated metal implants have been widely used over the past twenty years, with companies specializing in producing such devices for orthopaedic applications [84].

b) Coatings of TiO_2 and CaO_3

Nano TiO_2 coatings on Ti by spraying with magnetron equipment showed no better osteintegration than Ti with untreated surface. A deposition of $CaTiO_3$ with the deposition of deposited layer and crystallinity is also beneficial to bone mass formation [3].

VIII.1.2 Electro-chemical and Chemical Coatings

a) Hydroxyapatite deposits

Electrochemical treatments are commonly used to form a layer of HA on Ti [4]. Through electrochemical processes, HA layers with programmed morphology (plates, pins, particles) can be precipitated on a Ti substrate that can sometimes be heated to obtain a better deposition layer. For collagen immobilization, the same method for the deposition of β-TC_8 (tricalcium beta phosphate) may be used. For calcium Ti calcium phosphate

Materials Research Forum LLC
https://doi.org/10.21741/9781644900796

precipitation, low voltage currents have also been successfully used, as can be seen from the scanning electron microscopy in Figure VIII.1.

The technique is very useful for treating thin threads and fibers without the melting of titanium. Calcium nanoparticle calcium phosphate can be obtained under the Ti layer, using also acidic electrolytes. The obtained layer contains dehydrated bicalcium phosphate (55-85 nm) with a small amount of HA (20 / 25nm) but with an increase in the HA content with increasing the intensity of the current [38]; HA can also be deposited through an electrophoresis process by dynamically changing the voltage.

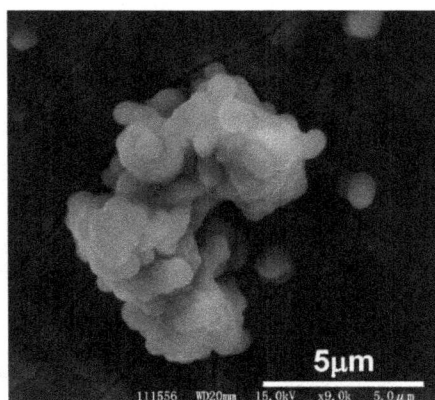

Figure VIII.1. Scanning electron microscopy of a calcium phosphate precipitate deposited on a titanium support [5].

b) Deposition of TiO_2

Tests for anodic oxidation and handling of Ti films in electrolytic solutions of Ca glyphosate or calcium acetate [6] have been performed. A porous nanotubular structure with self-organization was obtained by anodizing on Ti in Na_2SO_4 electrolytic solution containing NaF. The oxidation was performed at 20V with agitation of the electrolyte bath by ultrasonic or magnetic vibration. TiO_2 films in the form of nanotubes on Ti substrate were obtained using an electrochemical method. The formation and growth of a nanoscopic self-organizing layer can be achieved directly by anodizing electrolytes with NH_4. The diameter, length and thickness of the nanowires are significantly influenced by anodic conditions such as voltage, current density and anode time [39].

c) Micro-Arc Oxidation (MAO)

Micro-arc oxidation (also called plasma electrolytic oxidation or spark anode oxidation) is a convenient technique for forming oxide layers on metals. This technique has a good yield in the growth of TiO_2 films on the substrate Ti and TiO_2 on Zr substrate (Figure VIII.2.).

One of the advantages of the MAO technique is that the composite layer is porous and uniform on the surface of a complex metallic metal.

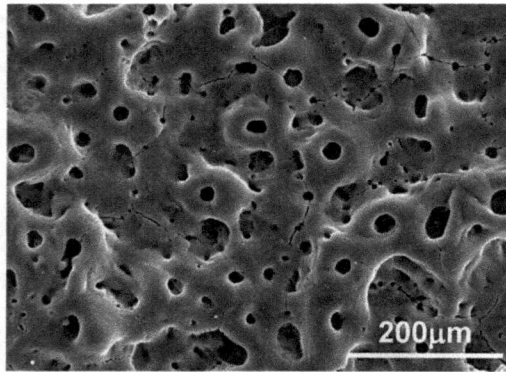

Figure VIII.2. SEM microscopy of a porous layer of ZrO_2 formed on Zr substrate by MAO [7].

Techniques of electrochemical anodic deposition and MAO are not very clearly separated. If a layer of oxide is formed with pores of connection with the metal substrate then this technique is considered as micro-arc oxidation. MAO is currently a technique used to produce thick or porous HA oxides.

d) Formation of surface modified layers by chemical treatment

The hard tissue compatibility can be improved by modifying the Ti surface instead of the HA coating layer. Numerous surface modification techniques have been developed without the use of HA or Ca phosphate. When Ca ions were implanted on the surface of an implant, the rate of precipitation of calcium phosphate increased by reducing by 2 days the formation of an osteoid tissue compared to the result obtained on a previously untreated implant [8]. This was based on substrate surface modification by implantation with ions. The calcium ion-modified surface showed a low percentage of Ca TiO_3 in the TiO_2 layer.

When titanium is immersed in alkaline solutions of NaOH or KOH, a hydrated Ti oxide containing 1 μm alkaline ions is formed on the Ti substrate [9]. After a heating step, the gel layer condenses, and its bonds with the substrate harden. When Ti is immersed in simulated body fluids, the alkaline ions are instantly released from the layer to the solution. At the same time, the hydronium ions are absorbed by the layer forming a thin layer of hydro-titanium, a gel that increases the supersaturation of the HA layer in the solution, near the surface. The gel acts to induce hydroxyapatite nucleation, and the HA layer will form much faster. This process is already marketed for an artificial hip fixation system.

Also, an acceleration of HA formation was reported with an immersion of the Ti substrate in the H_2O_2-containing $TaCl_3$ solution [10]. Also, the combination of acidic etching and alkaline treatment leads to the acceleration of HA deposition. A Ti oxide layer containing calcium hydroxide is always formed when Ti is immersed in solutions containing calcium; the oxide layer catalyses precipitation of calcium phosphate on titanium when immersed in Hank's solution. The fastest results of HA precipitation are immersion in an alkaline solution. Titanium can be chemically treated with a solution of H2O2 / HNO3 at 354K for $50 \div 60$ minutes for the TiO_2 and $CaTiO_3$ layer, then hydrothermal treated using an autoclave at 453 K for 24 minutes. This change promotes HA precipitation. Titanium is then introduced into a suspension of calcium hydroxide and heated, and its modification leads to the formation of a $CaTiO_3$ layer [11].

e) Inhibition of bone formation

When Ti-based alloys are used for bone fixation such as bone screws or nails implanted in bone marrow for orthopaedics, Ti alloys form a connective ligament (characterized by the rapid multiplication of young cells and their metaplasia as well as changes metabolic pathways of the old hematoma) and is sometimes assimilated to the bone. Therefore the bone can be fractured again when the fasteners are withdrawn after the healing of the bone, since Ti easily forms calcium phosphate on the surface [12]. For these reasons it is necessary that Ti-based alloys do not form this cloth for their safe use. The deposition of HA on metallic materials used for implants is a very good solution for integrating these materials into the human biological system.

Interaction between solid surfaces and biological systems is very important for many areas of medicine, technology and research. In general, only the surface of an implant is in direct contact with the host tissue and therefore this part of the material plays the central role in determining the biocompatibility of the materials used for implants. The surface of the material can change over time and far differently from the properties of the original material due to the oxidation and contamination phenomena. Even though the surface of an implant material plays an important role in the interactions between the implant and the cells, the relationship between surfaces and implants and the constituents of the biological tissue, and long-term integrity and chemical efficacy are little understood [13].

VIII.2 Powder Metallurgy

In the latest studies, a new technique is used, a wide variety of metallurgical products are obtained using a technology principally by classical technology, based on the casting of metal in semifinished fabrics or ingots that are subsequently for to plastic deformation [14-20]. This alternative is provided by powder metallurgy, where the products are obtained from metallic powders sometimes mixed with non-metallic powders and / or binders, by means of pressing and heating without melting (sintering). Obtaining parts through powder metallurgy is widely used for:

- the production of small and complicated parts whose obtaining by casting is not cost-effective either from the point of view of consumption (of materials, workmanship, etc.) nor from the point of view of precision;

- production of parts which, by their nature of operation, must have a porous structure;

- obtaining in pure state metals and alloys which are hardly fusible, materials in which melting and casting are not only difficult to achieve but lead to inadmissible impurities;

- obtaining materials whose components do not naturally alloy (pseudo alloys for W - Ag, W - Cu, Mo - Ag, Mo - Cu, graphite, graphite, bronze - graphite and cermets that associate in their structure a ductile metallic phase with a refractory and durable ceramic phase - oxides, carbides, etc.).

A tendency for powder metallurgy [15, 16, 21-30] (PM) is to use if to fabricate a new type of titanium (Ti) + magnesium (Mg) bioactive composite to enable stress-shielding reduction and obtain better biocompatibility compared with that of the traditional Ti and Ti alloys used for dental implants. Such composites are produced by well-known cost effective and widely used PM methods, which eliminate the need for complex and costly Ti casting used in traditional implant production. Magnesium (Mg) was discovered, which is a naturally occurring metallic element in the human body, and it is relatively affordable. It was researched in recent years as the basis for new types of biodegradable implants. Along with its very good biodegradation potential, Mg has a low E of 45 GPa, which is very similar to that of bones. However, Mg implants suffer from the difficulty to control their degradation rate after implantation. The implants corrode fast due to chloride ions that are present at high concentrations in the human body. The corrosion inhibits full recovery from injuries. Also, as pure Mg undergoes corrosion, toxic products are released, eventually leading to the necrosis of the surrounding tissue. Thus, previous studies on Mg as an implant material concentrated on Mg corrosion rates under exposure to bodily fluids and how the corrosion could be controlled. A few studies were aimed at boosting the bioactivity of Ti and Ti alloy implants by finishing their surface with Mg or Mg alloys.

Superior mechanical properties such as high strength are the major advantages of metals in biomedical applications [15, 16, 31-41]. But the prominent problem is that the young's modulus of most popular biomedical metal materials is relatively high comparing with human bones, which would cause "stress shielding" effect after implantation. Therefore, it is very attractive that the young's modulus of the Ti-Mg composites is much lower than the dominant metallic biomaterials and comparable with human bone. What's more, the strength of the Ti-Mg composites and the porous Ti is remarkably enhanced in comparison with those of the reported works due to the two-step sintering and the ultrasonic infiltration method, which is significant for the reason that the implant should have adequate strength before and after the degradation of Mg. All these features of the Ti-Mg composites demonstrate the promising prospect in load-bearing biomedical applications and potential to popularize. It must be mentioned that, although the Ti-Mg composites were designed to serve as a semi-degradable biomedical material, the degradation rate of the Mg matrix should be controlled according to the specific service condition. Further research on improving the corrosion resistance of the Ti-Mg composites is necessary, such as forming a dense layer between the Ti and Mg, which will be our next effort. Also, for the future all titanium alloy systems will be developed through powder metallurgy.

References

[1]   S. Sturbinger, C. Etter, M. Miskiewicz, F. Homann, B. Saldamli, M. Wieland, et al. Surface alterations of polished and sandblasted and acid-etched titanium implants after Er:YAG, carbondioxide, and diode laser irradiation. Int J Oral Maxillofac Implants 25 (2010) 104-111.

[2]   A. Pae, S.K. Kim, H.S. Kim, Y.H. Woo, Osteoblast-like cell attachment and proliferation on turned, blasted, and anodized titanium surfaces, Int J Oral Maxillofac Implants, 26(2011) 475-481.

[3]   Y. Oshida, J. Daly, Fatigue damage evaluation of shot peened high strength aluminum alloy. In: Meguid SA, editor. Surface engineering, NewYork, Elsevier Applied, 1990, pp.404-416. https://doi.org/10.1007/978-94-009-0773-7_42

[4]   A.T. De Wald, J.E. Rankin, M.R. Hill, M.J. Lee, H.L. Chen, Assessment of tensile residual stress mitigation in Alloy welds due to laser peening, J Eng Mater Tech, Trans ASME, 126 (2004) 81- 89. https://doi.org/10.1115/1.1789957

[5]   C.B. Dane, L.A. Hackel, J. Daly, J. Harrison, High power laser for peening of metals enabling production technology, Mater ManufacProcess, 15 (2000) 81-96. https://doi.org/10.1080/10426910008912974

[6]   S.A. Cho, S.K. Jung, A removal torque of the laser-treated titanium implants in rabbit tibia. Biomaterials, 24 (2003) 4859-4863. https://doi.org/10.1016/S0142-9612(03)00377-6

[7]   S. Hansson, K.N. Hansson, The effect of limited lateral resolution in the measurement of implant surface roughness: a computer simulation, J Biomed Mater Res A, 75 (2005) 472-477. https://doi.org/10.1002/jbm.a.30455

[8]   J.H. Park, R. Olivares-Navarrete, R.E. Baier, A.E. Meyer, R. Tannenbaum, B.D. Boyan, et al. Effect of cleaning and sterilization on titanium implant surface properties and cellular response. Acta Biomater, 8 (2012) 1966 -1975. https://doi.org/10.1016/j.actbio.2011.11.026

[9]   S. Ban, Y. Iwaya, H. Kono, H. Sato, Surface modification of titanium by etching in concentrated sulfuric acid, DentMater, 22 (2006) 1115-1120. https://doi.org/10.1016/j.dental.2005.09.007

[10] Y. Iwaya, M. Mchigashira, K. Kanbara, M. Miyamoto, K. Noguchi, Y. Izumi, et. al., Surface properties and biocompatibility of acid-etched titanium, Dent Mater J, 27 (2008) 415-421. https://doi.org/10.4012/dmj.27.415

[11] P.S. Vanzillotta, M.S. Sader, I.N. Bastos, G.A. Soares, Improvement of in vitro titanium bioactivity by three different surface treatments. DentMater, 22 (2006) 275-282. https://doi.org/10.1016/j.dental.2005.03.012

[12] K. Ungvarri, I.K. Pelsoczi, B. Kormos, A. Oszko, Z. Rakonczay, L. Kemeny, et al., Effects on titanium implant surfaces of chemical agents used for the treatment of peri implantitis, J Biomed Mater Res Part B:Appl Biomater, 94 (2010) 222- 229. https://doi.org/10.1002/jbm.b.31644

[13] U. Ingela, I.U. Petersson, J.E.L. Loberg, A.S. Fredriksson, E.K. Ahlberg, Semiconducting properties of titanium dioxide surfaces on titanium implants, Biomaterials, 30 (2009) 4471-4479. https://doi.org/10.1016/j.biomaterials.2009.05.042

[14] F.M. He, G.L. Yang, S.F. Zhao, Z.P. Cheng, Mechanical and histomorphometric

evaluations of rough titanium implants treated with hydrofluoric acid/nitric acid solution in rabbit tibia, Int J Oral Maxillofac Implants 26 (2011) 115-122.

[15] M. Baloga, A.M.H. Ibrahima, P. Krizika, O. Bajanaa, A. Klimovaa, et. all, Bioactive Ti + Mg composites fabricated by powder metallurgy: The relation between the microstructure and mechanical properties, Journal of the Mechanical Behavior of Biomedical Materials, 90 (2019) 45–53. https://doi.org/10.1016/j.jmbbm.2018.10.008

[16] S. Jiang, L.J. Huang, Q. An, L. Geng, X.J. Wang, S. Wang, Study on titanium-magnesium composites with bicontinuous structure fabricated by powder metallurgy and ultrasonic infiltration, Journal of the Mechanical Behavior of Biomedical Materials, 81 (2018) 10–15. https://doi.org/10.1016/j.jmbbm.2018.02.017

[17] P. Moshayedi, G. Ng, J.C. Kwok, G.S. Yeo, C.E. Bryant, J.W. Fawcett, K. Franze, et all, The relationship between glial cell mechanosensitivity and foreign bodyreactions in the central nervous system. Biomaterials, 35 (2014) 3919–3925. https://doi.org/10.1016/j.biomaterials.2014.01.038

[18] J. Jain, P. Cizek, K. Hariharan, Transmission electron microscopy investigation ondislocation bands in pure Mg. Scr. Mater, 130 (2017) 133–137. https://doi.org/10.1016/j.scriptamat.2016.11.035

[19] R. Narayan, ASM Handbook, Volume 23, Materials for Medical Devices. ASMInternational, Materials Park, 2012. https://doi.org/10.31399/asm.hb.v23.9781627081986

[20] M. Niinomi, M. Nakai, J. Hieda, Development of new metallic alloys for biomedicalapplications. Acta Biomater., 8 (2012) 3888–3903. https://doi.org/10.1016/j.actbio.2012.06.037

[21] B. Piotrowski, A.A. Baptista, E. Patoor, P. Bravetti, P. Laheurte, Interaction ofbone–dental implant with new ultra-low modulus alloy using a numerical approach.Mater. Sci. Eng. C, 38 (2014) 151–160. https://doi.org/10.1016/j.msec.2014.01.048

[22] J.I. Qazi, H.J. Rack, Metastable beta titanium alloys for orthopedic applications.Adv. Eng. Mater., 7 (2005) 993–998. https://doi.org/10.1002/adem.200500060

[23] M. Qian, F.H. Froes, Titanium Powder Metallurgy: Science, Technology andApplications. Elsevier Science (ISBN: 9780128009109), 2015.

[24] A. Revathi, A.D. Borras, A.I. Munoz, C. Richard, G. Manivasagam, Degradation mechanisms and future challenges of titanium and its alloys for dental implant applications in oral environment. Mater. Sci. Eng. C, 76 (2017) 1354–1368. https://doi.org/10.1016/j.msec.2017.02.159

[25] P. Sevilla, C. Sandino, M. Arciniegas, J. Martinez-Gomis, M. Peraire, F.J. Gil, Evaluating mechanical properties and degradation of YTZP dental implants. Mater.Sci. Eng. C, 30 (2010) 14–19. https://doi.org/10.1016/j.msec.2009.08.002

[26] G. Song, Control of biodegradation of biocompatible magnesium alloys. Corros.Sci., 49 (2007) 1696–1701. https://doi.org/10.1016/j.corsci.2007.01.001

[27] M.P. Staiger, A.M. Pietak, J. Huadmai, G. Dias, Magnesium and its alloys asorthopedic biomaterials: a review. Biomaterials, 27 (2006) 1728–1734. https://doi.org/10.1016/j.biomaterials.2005.10.003

[28] N. Taniguchi, S. Fujibayashi, M. Takemoto, K. Sasaki, B. Otsuki, T. Nakamura, T. Matsushita, T. Kokubo, S. Matsuda, Effect of pore size on bone ingrowthinto porous titanium implants fabricated by additive manufacturing: an in vivo experiment.Mater. Sci. Eng. C, 59 (2016) 690–701. https://doi.org/10.1016/j.msec.2015.10.069

[29] Y.L. Xi, D.L. Chai, W.X. Zhang, J.E. Zhou, Titanium alloy reinforced magnesium matrix composite with improved mechanical properties. Scr. Mater., 54 (2006) 19–23. https://doi.org/10.1016/j.scriptamat.2005.09.020

[30] Y.C. Xin, K.F. Huo, H. Tao, G.Y. Tang, P.K. Chu, Influence of aggressive ions on the degradation behavior of biomedical magnesium alloy in physiological environment.Acta Biomater., 4 (2008) 2008–2015. https://doi.org/10.1016/j.actbio.2008.05.014

[31] R. Zeng, W. Dietzel, F. Witte, N. Hort, C. Blawert, Progress and challenge formagnesium alloys as biomaterials. Adv. Eng. Mater., 10 (2008) 3–14. https://doi.org/10.1002/adem.200800035

[32] C. Zhao, X. Zhang, P. Cao, Mechanical and electrochemical characterization ofTi–12Mo–5Zr alloy for biomedical application. J. Alloy Compd., 509 (2011) 8235–8238. https://doi.org/10.1016/j.jallcom.2011.05.090

[33] M. Balog, M. Snajdar, P. Krizik, Z. Schauperl, Z. Stanec, A. Catic, Titanium-magnesium composite for dental implants (BIACOM). In: TMS 2017 146th Annual Meeting & Exhibition Supplemental Proceedings, Springer, 2017, pp. 271–284. https://doi.org/10.1007/978-3-319-51493-2_26

[33] D.A. Basha, H. Somekawa, A. Singh, Crack propagation along grain boundaries and twins in Mg and Mg–0.3 at%Y alloy during in-situ straining in transmission electron microscope. Scr. Mater., 142 (2018) 50–54. https://doi.org/10.1016/j.scriptamat.2017.08.023

[34] B. Chang, W. Song, T. Han, J. Yan, F. Li, L. Zhao, H. Kou, Y. Zhang, Influence of pore size of porous titanium fabricated by vacuum diffusion bonding of titanium meshes on cell penetration and bone ingrowth. Acta Biomater., 33 (2016) 311–321.

https://doi.org/10.1016/j.actbio.2016.01.022

[35] Q. Chen, G.A. Thouas, Metallic implant biomaterials. Mater. Sci. Eng. R. Rep., 87 (2015) 1–57. https://doi.org/10.1016/j.mser.2014.10.001

[36] C.D. Dunand, Processing of titanium foams. Adv. Eng. Mater. 6 (6), 2004, pp. 369–376. https://doi.org/10.1002/adem.200405576

[37] Z. Esen, B. Dikici, O. Duygulu, A.F. Dericioglu, Titanium–magnesium based composites: mechanical properties and in-vitro corrosion response in Ringer's solution. Mater. Sci. Eng.: A, 573 (2013) 119–126. https://doi.org/10.1016/j.msea.2013.02.040

[38] Z. Esen, Z., Bütev, E., Karakaş, M.S., 2016. A comparative study on biodegradation and mechanical properties of pressureless infiltrated Ti/Ti6Al4V–Mg composites. J. Mech. Behav.Biomed. Mater. 63, 273–286.
https://doi.org/10.1016/j.jmbbm.2016.06.026

[39] Z. Fan, B. Zhang, Y. Gao, X. Guan, P. Xu, Deformation mechanisms of spherical cell porous aluminum under quasi-static compression. Scr. Mater., 142 (2018) 32–35. https://doi.org/10.1016/j.scriptamat.2017.08.019

[40] M. Geetha, A.K. Singh, R. Asokamani, A.K. Gogia, Ti based biomaterials, the ultimate choice for orthopaedic implants – a review. Progress. Mater. Sci. 54 (3), 2009, pp. 397–425. https://doi.org/10.1016/j.pmatsci.2008.06.004

[41] Z. Esen, Ş. Bor, Processing of titanium foams using magnesium spacer particles. Scr.Mater. 56 (5), 2007, pp. 341–344. https://doi.org/10.1016/j.scriptamat.2006.11.010

Keyword Index

About the Authors

Petrică VIZUREANU
Professor Ph.D., M.Sc., Eng.
Director of the Department of Technologies and Equipment for Materials Processing
"Gheorghe Asachi" Technical University of Iasi, Faculty of Materials Science & Engineering
Email: peviz2002@yahoo.com
Website: www.afir.org.ro/peviz/

Full Professor with a teaching experience of over 25 years, a very rich experience in project management of national and international research projects (Director – 6, member – 35), concerns concretized in many articles in different competence areas: Biomaterials (characterization, testing and material expertise; medical devices; surface properties; tissue-implant interaction phenomena), materials science, unconventional energy, refractory materials, computer assisted design, geopolymers, safety and health at work, management and commercial engineering.

Publications: over 150 ISI/WoS articles published in journals and conference proceedings, books and chapters in specialized books – **Author** of 17 national books and 9 international chapters/books., **Editor** of books INTECH OPEN (4 books + 6 chapters), H-index – 16 (>500 citations for 126 articles in Web of Science / 110 articles in SCOPUS).

Editor-in-Chief: European Journal of Materials Science and Engineering, indexed by DOAJ, CAS and CiteFActor
Guest Editor: MATERIALS (MDPI): Advanced Surface Treatment Technologies for Metallic Alloys; COATINGS (MDPI): Surface Treatment of Metals.
Chairman: ICIR EUROINVENT - International Conference on Innovative Research (Scopus & Web of Science Proceedings) 2015-2020.

Titanium-Based Alloys for Biomedical Applications
Materials Research Foundations **74** (2020)

Materials Research Forum LLC
https://doi.org/10.21741/9781644900796

Mădălina Simon BĂLȚATU
Lecturer Ph.D. Eng.
"Gheorghe Asachi" Technical University of Iasi, Faculty of Materials Science &
Engineering
Email: cercel.msimona@yahoo.com
Website: www.afir.org.ro/msb/

Researcher and lecturer at "Gheorghe Asachi" Technical University of Iasi, Faculty of Materials Science & Engineering. Most of the author's research activity achievements are in the field of biomaterials being involved in preliminary studies for the foundation and development of new biocompatible materials based on titanium, zirconium, cobalt and magnesium alloys. Her publications focus on new biomaterials and advanced characterization of various compounds regarding their morphological, physical/chemical a respectively their behaviour. As a consequence of the scientific and research activity carried out in the biomaterials field, a number of over 22 articles have been published, of which 5 are indexed in journals with an impact factor > 1.00, 1 book and 1 international chapter book. She is involved in Organizing Committee of European Exhibition of Creativity and Innovation – EUROINVENT, and International Conference on Innovative Research, since 2015, which gathers more events: invention and research projects exhibition, conference on innovative research and book and art salon. She also is on the Committee Board of the European Journal of Materials Science and Engineering (www.ejmse.tuiasi.ro) and carries out activities in 2 national projects as director and in 4 projects as member (2 national and 2 international).

www.ingramcontent.com/pod-product-compliance
Lightning Source LLC
Chambersburg PA
CBHW070729220326
41598CB00024BA/3361